Productivity Theory for Industrial Engineering

Systems Innovation Series

Adedeji B. Badiru

Air Force Institute of Technology (AFIT) – Dayton, OH

PUBLISHED TITLES

Productivity Theory for Industrial Engineering

Ryspek Usubamatov

CRC Press
Taylor & Francis Group
Boca Raton London New York

CRC Press is an imprint of the
Taylor & Francis Group, an **informa** business

CRC Press
Taylor & Francis Group
6000 Broken Sound Parkway NW, Suite 300
Boca Raton, FL 33487-2742

First issued in paperback 2021

ISBN-13: 978-0-367-78119-4 (pbk)
ISBN-13: 978-1-138-48321-7 (hbk)

Visit the Taylor & Francis Web site at
http://www.taylorandfrancis.com

and the CRC Press Web site at
http://www.crcpress.com

To the Readers

*With the hope that this work will help to solve productivity
problems in manufacturing and industrial engineering.*

Contents

Preface

Throughout the centuries, manufacturing industries accumulated experience in creating different machines for production of various goods, work parts and products. All these diversities of machines and systems with different designs are solving pivoted problems of economics to increase the productivity rate of manufacturing processes with high quality of products. The productivity rate of manufacturing systems is the primary area of any industry in the world. Manufacturers are interested in any publication that focus on the problem of solving the productivity rate problems. Macro and microeconomics consider two types of productivity. In macroeconomics, the term productivity is the ratio of output to input, which is the efficiency of the economic system according to the fundamental science. Microeconomics considers physical productivity as the number of products fabricated per given time as accepted by The American Society of Mechanical Engineers (ASME). Productivity rate is one important indicator criteria for an industry engineer to improve the manufacturing system and provide a good sustainable output of production processes.

The productivity theory for industrial engineering is not new, and there are numerous publications with fundamental approaches and mathematical models of physical productivity for different designs of industrial machines and systems. However, known mathematical models for the manufacturing systems are simplified with numerous assumptions that are not well developed analytically and lead to severe errors in calculations. The modern industrial machines and systems are considered as systems that contain different mechanisms and arrangements, and represent conglomerates of mechanical, electrical and electronic units. The structures of manufacturing machines and systems are also having large variances with serial, parallel and serial–parallel arrangements of machines disposed according to the technological process for machining products. Any failure of components leads to downtime of expensive production systems that should be minimised by different solutions. Manufacturers are correcting the results by data of practical experience in the designs of productive machines and systems and optimisation of the machining process. This is the reason that industries need productivity theory with mathematical models that combine all aspects of production processes and enables solving numerous engineering problems. This is an important position that enables predicting the design and output of new manufacturing machine and systems in preliminary stages. Manufacturers need correct and clear mathematical models that enable computing with high accuracy the productivity rate of machines and systems.

Proposed productivity theory for industrial engineering collected all achievements in analytical and practical experience in formulating the physical productivity for different manufacturing systems with simple and complex designs. Productivity theory for industrial engineering discovers laws and reasonable links for creating manufacturing machines and systems. It is necessary to point out that this productivity theory is universal and can be applied to any type of industries, i.e. manufacturing, textile, transport, chemical, etc. The presented productivity theory maximally avoids simplifications in analysis and mathematical models of productivity rate for different designs and structures of the manufacturing systems. Renovated and corrected productivity theory enables accurate solving of the problems of design the machines and systems based on innovative technologies with high manufacturing productivity rates. Productivity theory enables defining optimal balancing technological processes, optimal machining regimes and optimal structures of manufacturing machines and systems by criterion of necessary or maximal productivity rate and sustainable improvement of production process.

The fundamentals of productivity theory for industrial engineering are based on several engineering and mathematical theories that include theories of technological processes, machine designs, reliability, probability, optimisations, engineering management, economics and others. Combined application of these theories enables the development of a holistic productivity theory for industrial engineering. Productivity theory considers all factors that influence the output and design of manufacturing machines and systems. The first factor is the technological process that is the essence of design for any manufacturing machine and system, which defines the limits of the productivity rate and time spent for production process. Other theories applied to design of manufacturing machines and systems enable a decrease in the non-productive or auxiliary and idle times of production processes. The objective of productivity theory is to provide the methods and mathematical models for studying the regularities and reasonable links in design of manufacturing machines and systems with optimal structures and maximal productivity rate. In addition, this theory enables demonstrating the productivity losses and their reason in real production environments. Solutions of these problems are making the progress in developing industrial machines based on the application of innovative technological processes that enable increase in their productivity rate. Such universality of productivity theory for solutions of productivity problems for manufacturing machines and systems can be used in complex and purposeful approaches for sustainable improvement of production process. Hence, the perspective designs of manufacturing machines and systems request from engineers and technologists the wide knowledge and prospects and understanding the essence and regularities in the development of industrial engineering. Productivity theory for industrial engineering can be a core course for manufacturing and industrial engineering departments and faculties of universities and will be desk book for practitioners and manufacturers.

Nomenclature

A	availability of a machine (reliability factor)
C_e	expenses for production processes
C_f	fixed capital
C_s	service capital
C	empirical constant resulting from regression analysis of the tool life
E	efficiency of production system
$F(x)$	cumulative distribution function
T	cycle time for the machining of one product
T_c	tool life of a cutter
I	probability of downtime
L	labour productivity
M	total number of manufactured products
N	service years for the economic system; the capacity of the buffer
N_l	limited capacity of the buffer
$P(t)$	failure probability
Q	productivity of a manufacturing system machine tool, etc.
Q_a	actual productivity of a machine
Q_c	cyclic productivity of a machine
Q_m	machine productivity
Q_{max}	maximal productivity rate
Q_t	technological productivity of a machine
$R_i(t)$	is the probability that workstation, section or some unit i will work during a defined time (t) without failures
S	salary of employees
V	cutting speed of a machining process
V_n	normative cutting speed
a	total number of machine downtimes
b	empirical constant that depends on the cutter tool material
c	coefficient change in machining time, speed, velocity, etc.
f_c	correction factor for machining time
f_m	management factor

$f_{s.i}$	correction factor for sections due to reliability
f_s	correction factor for probabilities
$f(x)$	probability density distribution of random event
m_w	mean time to work between two failures
m_r	mean time to repair
n	number of machine downtimes; number of tool replacement
q	number of serial workstations, machine tools
q_s	number of all workstations in service by one technician
p	number of parallel products
p_s	number of parallel workstations, machine tools
p_τ	number of workstations located on the transport sector τ of the transport rotor
p_γ	number of workstations located on the idle sector γ of the working rotor
t_a	auxiliary time
t_{av}	average machining time
t_m	machining time
t_{mb}	machining time at the bottleneck workstation
t_{mo}	total machining time
t_{mn}	normative machining time (min/part)
z	number of products machined per observation time
θ	observation time
θ_a	auxiliary time for the production process
θ_{cs}	idle time due to control system
θ_m	idle time due to managerial problem
θ_i	idle time of a machine, workstation
θ_p	production process time
θ_r	repair time of a machine
θ_w	machine work time
θ_t/z	idle time referred to one product due to technical problems of a machine, (min/work part)
θ_{org}/z	idle time referred to one product due to managerial and organisational problems, (min/work part)
λ	failure rate
λ_{cs}	failure rate of control system
λ_s	failure rate of workstation or machine
λ_{sb}	failure rate of the bottleneck machine tool or workstation
λ_{tr}	failure rate of transport with the workpiece feed mechanism
λ_{bf}	failure rate of a buffer
λ_b	failure rate of a bottleneck section
λ_i	failure rate of i mechanisms (gears, shafts, bearings, etc.)
$\Delta\lambda_s$	added failure rate for a bottleneck section
Δt	displacement time of the sequence station work
ΔQ	productivity rate losses

Author

Ryspek Usubamatov graduated from Bauman Moscow State Technical University. He is a professional engineer in mechanical, manufacturing and industrial engineering. He obtained his Ph.D. in 1972 and doctor of technical sciences in 1993. He worked as an engineer at a company and as a lecturer in Kyrgyzstan and Malaysian universities. He has supervised around 100 professional engineering students, 15 M.Sc. and 7 Ph.D. His areas of research include productivity theory for industrial engineering, gyroscope theory and wind turbines. He has published 7 books, 30 brochures, more than 300 manuscripts and 60 patents of inventions.

chapter one

Introduction to productivity theory for industrial engineering

Industrial engineering encompasses various technologies and production machineries that focus on high productivity and quality of fabricated goods. The main purpose underlying any industrial process is to strive for productivity and profitability. Industries use several mathematical models for productivity, which do not allow a production system to be described correctly. Exact mathematical models for the productivity rate include parameters such as technologies, design, reliability of machineries and management systems, which are the crucial problems in engineering. Modern manufacturing processes implement different machineries at preparatory, machining and assembly stages, where machining remains the most important one. Contemporary tendency of a manufacturing process is complication and automation of machines and systems, which is a natural result of engineering evolution. Production systems have become increasingly complex in design, which in turn are faced with reliability problems and is one of the most important attributes in all machineries and plays a key role in the cost-effectiveness of an industrial system. The productivity rate of machines and systems has a direct relationship with the reliability of their units, which should be described analytically. Analysis and mathematical modelling of machine productivity makes it possible to embed standard attributes of reliability as components of productivity rate equations for different designs and structures of industrial machines and systems.

1.1 Introduction

The term industrial engineering comprises a wide area of engineering that deals with the optimisation of complex processes and systems, and leads to eliminate wastage of production time, decrease machine time and decrease energy consumption and other resources that do not generate value. In addition, industrial engineers figure out how to better engineering processes and systems that improve quality and productivity. Industrial engineering encompasses a range of specialised fields of engineering, particularly areas of technologies and production machineries that focus on high output with high quality. This is a main part of economic activity of industrial companies that represents the critical success factor of production systems. To define optimal engineering solutions and to describe mathematical models for productivity of manufacturing machines and systems, the concepts involved in it are primary important problems in production systems.

The main purpose underlying any production process is to strive for productivity and profitability, which are typically specified concepts of efficiency. Production is a system of combining various technological processes, information, machinery and material inputs of production to manufacture products for consumption. All production systems are processes that transform resources into useful goods and are based on application of labour, equipment, materials, land, buildings, etc. The technological processes constitute the works in the preparatory, machining and assembly stages of manufacture, and finally the finished goods characterise a typical production system. The technological processes subject to the constraints of the capacity of the production system, which are limited by the nature of physics of technology and quality of products, limit the system's ability to meet output expectations. The management of technological processes, i.e. the planning and control of the production system to achieve acceptable outputs, is an important task of the production manager.

The productivity of the manufacturing system and the quality of products are major factors in determining the efficiency of production systems. Some researchers have deliberated productivity as a phenomenon in production processes. They suggest that the measurement of productivity in production processes shall be developed so that it will indicate an increase or decrease in the productivity of the company. Researchers regard the measurement of productivity rate as an important part of the productivity phenomenon and solve the problems related to measuring it at a long time.

However, this approach is indistinct, indefinite, does not have links and does not give clear picture dependency of productivity on technological parameters of processes, design of manufacturing machine and systems.

In industries, there are several productivity measurement models. Finding an appropriate model for production systems is crucial and an important engineering and economic problem. First of all, it should be described as the main technological process of production systems, and after that study in detail must focus on the most interesting processes from the point of view of productivity and the solutions to measure such processes. Productivity should be identified in connection with profitability and then the processes and machinery-generating products.

Analysis of known models for evaluation of the productivity for production processes demonstrates that all of them consider productivity in terms of economics and far from engineering terms. Methodologies for calculating system productivity have been represented in several key publications that consider all aspects of the production process. For example, overall equipment effectiveness is a well-known concept in maintenance and is a way of measuring the effectiveness of a machine and systems. It is the backbone of many techniques employed in asset management programs. However, known models for productivity of different industrial machines and systems are represented by generalised equations and approaches in terms of economics, which do not allow one to discover all engineering properties of the production system. Real process of a company generates the production output, and mathematical models of the production processes can describe it. It refers to a series of components in production processes, which include parameters of technologies, design and reliability of machineries, management systems, that ensure the quantity and quality of products. Productivity index is a primary factor in economics that has two broad branches: macroeconomics and microeconomics.

Macroeconomics studies the economy as a whole. It also studies the country's economy or the global economy and the economic decisions of a firm or groups of firms. In addition, macroeconomic problems are specific to an industry or a commodity. Macroeconomic productivity is a measure of the efficiency of production systems, where the term productivity is a ratio of the output to the input of labour or the amount of output per unit of input (labour, equipment, energy and capital). There are many different ways of measuring economic productivity. Productivity may be conceived of as a metric of the technical or engineering efficiency of production and might be measured based on the number of hours it takes to produce a good. Economists provide the definition of the efficiency of production as the ability to accomplish a job with minimum expenditure of time and effort. In addition, in economics, there are several partial mathematical models and definitions of productivity and efficiency for production systems. With the help of production function, it is possible to describe simply the mechanism of economic growth that presents a production increase achieved by an economic system. Economic growth is created by two factors: an increase in production input and an increase in production output. These components are represented mathematically by the following typical equation for the productivity of economic system [8]:

$$L = \frac{M}{C_e} = \frac{QN}{C_f + (C_s + S)N} \tag{1.1}$$

where L is the productivity economic system; M is the total number of manufactured products (output); C_e is the expenses for production processes (input); C_f is the fixed capital (cost of buildings, roads, lorries, facilities, machinery, power station, computer centre, etc.); C_s is the service capital (cost of materials, energy, maintenance, cooling, water supply, etc.); S is the salary of employees, N is the service years for the economic system (factory, plant, industry, etc.); Q is the physical productivity rate of the economic system (products/year).

Economists analysed Eq. (1.1) and demonstrated how productivity of economic system depends on the value of its components and their influence on the economic system. Increase in the physical productivity Q gives 60%–70% of economic efficiency (automation and new technology), decrease in the fixed capital C_f and the service capital C_s gives 15%–20% of economic efficiency as quality of products, and decrease in the number of employees S gives 10%–15% of economic efficiency. Therefore, physical productivity is the main and weighed component in the efficiency of an economic system.

Growth potentials in productivity of economic system vary greatly by industry, and as a whole, they are directly proportionate to the engineering development in the branch. New and fast-developing industries attain stronger growth in the productivity of economic system. Developed industries demonstrate productivity growth in small steps. By the accurate measurement of productivity of economic system, it is possible to appreciate these small changes and create an organisation culture, where continuous improvement is a common value. Today, experts understand that human and social capitals together with competition have a significant impact on productivity growth. Figure 1.1 depicts the percentage of average impact of parameters for economic efficiency of production systems, i.e. increase in the physical or machine productivity, decrease in the fixed and service capitals and decrease in the number of employees.

Microeconomics studies the economic decisions of an individual firm or groups of firms. Microeconomics is considered mainly as the index

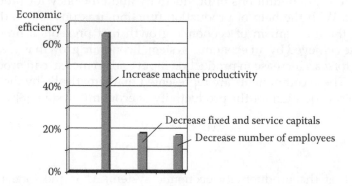

Figure 1.1 Growth of economic efficiency versus increase in the machine productivity, decrease in the fixed and service capitals and decrease in the number of employees.

of physical or machine productivity, where the term productivity refers to the ratio of the number of products fabricated per time. Engineering science (ASME) presents the term of productivity as the amount of work that can be accomplished in a given time period. The amount of work is represented as the number of products that can be discrete and continuous for production systems in the form of number of products, weights, length, volume, etc., which depends on the type of industries. Dimensions for physical or machine productivity represent the number of products/ time, metre/time, litre or m³/time, kg/time, etc. Differences in productivity measurement models are made transparent and evaluated by clear technical units in productivity theory and can be used by the following economic models of productivity, which is the highest level of economics.

Physical productivity of machines and systems is important and is a multifactor index that depends on the technology of processes, the reliability of mechanisms, machines and units, and indices of managerial and organisational activities, which definitely affect the output of production systems. All these components can be represented in mathematical models for productivity rate of machines and systems. Manufacturers need the mathematical models for the productivity rate for different designs and complexity of industrial machines and systems. Such mathematical models should base on the reasonable links between all parameters that play an important role in development and evaluation of the production systems. In addition, these mathematical models should reflect changes in the production processes. It means mathematical models for productivity rate of industrial machines and systems are flexible tools that can give the ability in calculating the economic indices of production systems and take fast decision for micro and macroeconomics policies. Hence, the productivity theory for industrial machines and systems plays a primary role in the efficiency of the economic system.

The object of productivity theory is analysis of regularities of development for industrial engineering and definition of reasonable links and dependencies, and their quantitative and qualitative descriptions. These positions include analysis of factors that affect the productivity rate of industrial machines and systems. Productivity theory contains mathematical models for quantitative dependency of productivity rates of industrial machines and systems on their design, structural and exploitation parameters, and their reliability. Qualitative indices of products are a subject of technology theory and are considered by the productivity theory from the point of view to increase the productivity rates of machines and systems. Nevertheless, any kind of technological processes have restrictions due to physical processes that cannot give a big change in the productivity rate. Increase in the productivity rate of a machine or system is possible by the new technological processes with high physical ability in the intensification of the processes. However, the high potential of new technologies can be implemented on the new construction of machines and systems. Practice demonstrates that any attempt to use new technologies on the old machines and systems does not give expected results. Productivity theory with analytical approaches enable solving these problems optimally based on the criterion of high productivity rate for industrial machines based on applied technology and reliability of the system.

The productivity theory for industrial machines and systems is not new; there are key publications with fundamental approaches and mathematical models of physical productivity for different designs of industrial machines and systems [16,17]. However, known mathematical models for the manufacturing systems are simplified with numerous assumptions and not well developed analytically, which leads to severe errors in calculations. This is the reason that known mathematical models of productivity theory for industrial machines and systems need deep revisions and corrections. Renovated and corrected productivity theory considered new mathematical models for productivity rate of industrial machines and systems, described accurately manufacturing processes and met with the requirements of modern production systems and economics.

1.2   Production-based industries, machineries and technologies for manufacturing systems

Since the Industrial Revolution, production-based industries clearly demonstrate the evolution of technologies and machineries. Contemporary automated lines, computerised workshops and factories with industrial robots are the result of scientific and engineering work.

From known types of industries, this book mainly considers production-based ones and derived that the productivity theory is applied to the primary and secondary industries. Primary industry is described by the main activity, which is extraction of raw materials from natural products untreated by people, from the land or sea, and considering mining, quarrying, forestry, farming and fishing. Secondary industry is described by the manufacture or the process of raw materials into consumable goods.

Manufacturing is the production of goods for use and based on use of labour and machinery with metallurgical, casting, forging, chemical, textile, biological, etc., processing. The term may refer to a wide range of human activity, from handicraft to high-tech, but most commonly applied to industrial production. Such finished goods may be used for manufacturing other, more complex products, such as aircraft, household appliances or automobiles and numerous other products. Modern manufacturing includes all intermediate processes required for the production and integration of a product's components. Typical manufacturing processes or technologies are material removal, material forming, additive processes, joining, assembling, packaging, non-traditional technologies, etc., and implemented in manufacturing systems.

A manufacturing system can be defined as the arrangement and operation of machines, tools, material, people and information to produce a value-added physical, informational or service product, whose success and cost is characterised by measurable parameters. All manufacturing or production systems and processes are classified by the following three common types: mass, batch with internal divisions (small batch, medium batch, large batch productions) and job shop. In mass production, equipment are specialised in execution identical, repeated operations of technological process.

In batch production, equipment are specialised in execution of two or more fixed operations, interchanged in defined sequences. In job-shop production, equipment is loaded by different works and does not have stable fixed operations. Today some industries have new types of manufacturing system that calls as mass and fast changeable production system that have properties of mass and batch productions. Manufacturing is usually directed towards the mass production of products for sale to consumers at a profit. Modern manufacturing includes all intermediate processes required for the production and integration of a product's components. The manufacturing sector is closely connected with engineering and industrial design.

These properties of the production processes reflect on the designs of the manufacturing facilities. Analysis of the manufacturing machines and systems for different productions depicts the following picture. In the mass production system, items to be processed flow through a series of steps, or operations, that are common to most other products being processed. Since large volumes of throughput are expected, specially designed equipment and methods are often used so that lower production costs can be achieved. Frequently, the tasks handled by workers are divided into relatively small segments that can be quickly mastered and efficiently performed. Examples include systems for assembling automobile engines and automobiles themselves, as well as other consumer products such as televisions, washing machines and personal computers. Mass production systems are often referred to as assembly systems or assembly line systems and, as noted later, are common in mass production operations. Mass production system is equipped by the machine tools and automated lines with fixed designs.

In the batch system, general-purpose equipment and methods are used to produce small quantities of output (goods or services) with specifications that vary greatly from one batch to the next. A given quantity of a product is moved as a batch through one or more steps, and the total volume emerges simultaneously at the end of the production cycle. Examples include systems for producing specialised machine tools or heavy-duty construction equipment, specialty chemicals and processed food products, or in the service sector, the system for processing claims in a large insurance company. Small batch production systems are often referred to as job shops.

Batch production systems are base for the flexible manufacturing systems with the computerised numerical control (CNC) machine tools. The batch production system mentioned thus far is often found in combination with other types.

The third type of production system is the project, or job-shop system, i.e. for a single, one-of-a-kind product. Because of the singular nature of project systems, special methods of management have been developed to contain the costs of production within reasonable levels. Job-shop production is equipped mainly by manually controlled machine tools and can have same properties and machinery with small batch production system. Analysis of the machinery designs for different types of production systems demonstrates that each type of production equipped with own designs of machinery cannot be used effectively at other systems.

Technology. One important aspect of manufacturing process is technology. Industrial technology is the use of engineering and manufacturing technology to make production faster, simpler and more efficient. An industrial technologist involves the processes of management, operation and maintenance of complex operation systems. Manufacturing systems comprise production technologies and associated services, processes, plants and equipment, including automation, robotics, measurement systems, cognitive information processing, signal processing and production control by high-speed information and communication systems.

Advanced manufacturing systems involve manufacturing operations that create high-tech products, use innovative techniques in manufacturing and invention of new processes and technologies for future manufacturing. This competitive advantage can only be built on manufacturing strength, which is best achieved through the acquisition of the highest technology and equipment in time. Those who master advanced manufacturing technologies will be endowed with the right machinery to respond to needs related to cost, quality and cycle time.

Today, industrial countries are homeland to major and advanced manufacturing processes. These manufacturing industries provide technologies and solutions, which are needed to respond to major challenges of the nearest future to ensure a better future for society. Advanced manufacturing systems integrate different technologies and knowledge into manufacturing process, which help to optimise the production and factories. An industrial technology professional uses engineering and fabrication skills to produce a streamlined manufacturing process. These systems comprise not a single technology but rather a combination of different technologies which include, among others, material engineering technologies (e.g. metal forming, metal removing, knitting, chipping, assembling, etc.), electronic and computing technologies, and their combination, measuring technologies (including optical and chemical technologies), transportation technologies and other logistic technologies. The industrial technology field employs creative and technically proficient individuals who can help a company achieve efficient and profitable productivity.

Machinery. Since the Industrial Revolution, manufacturing machines passed the evolution process, starting from simple designs and today represented manufacturing systems as advanced in designs and automations like automated lines and factories. This variety of manufacturing machines enables demonstrating their evolution, direction of development and the change in increase of productivity.

Industries use two types of machinery that should be separated for the following analysis. The first type is the power-related machinery that converts electric, thermal, chemical, etc., into mechanical energy. The second type is the general-purpose machinery that is used in production systems to manufacture goods for consumers. There is a special-purpose machinery that is machinery designed for exclusive use in one or a small cluster of activities. The general- and special-purpose machinery are widely used in production processes of different type goods in manufacturing industries. This type of machinery is represented in metal-forming and metal-removing technologies that accounted more value added in the machinery and equipment-manufacturing sector.

Manufacturing technology provides the machinery that enable production of all manufactured goods. Machinery includes machine tools, automated lines with machine tools that are called workstations and other related equipment and their accessories and tooling used to perform specific operations on manufactured materials to produce durable goods or components. Related technologies include computer-aided design and computer-aided manufacturing as well as assembly and test systems to create a final product or subassembly. In addition, big variances in constructional decisions of manufacturing machines and systems give rise to big variances in designations of machineries. For example, some publications represent manufacturing lines as a flow line, transfer line, conveyor line, etc. Productivity theory uses the following designations for consistency: manufacturing system is an arrangement created by machine tools of independent work and equipped with simple transporters for work parts by rolled tables, chutes, etc.; automated line is an arrangement created by machine tools or workstations of hard joining and equipped with transport and control systems. These manufacturing arrangements can be different designs and structures, but with independent or dependent work of basic technological machineries.

Technologies are basis for machinery designs. Manufacturing special machinery and automated machinery is the backbone of production systems, because automation in solving primary problems increases productivity. Regular practice demonstrates that upgrading or rebuilding existing automation systems or special machines is the most productive and cost-effective solution.

Analysis of machinery's designs for all type of industries demonstrates that the basic structure of any type of industrial machines is the same and can be presented as shown in Figure 1.2. Structural difference of machines is in executive mechanisms that form the machine's designs. The properties of executive mechanisms play important role in productivity theory for industrial machines and systems. Executive mechanisms implement machining, auxiliary motions and control the work of them. Reliability of the executive mechanisms is a primary factor presented by attributes in mathematical models of productivity rate for industrial machines and systems.

Metal-forming technologies. Metal-forming processes belong to the preparatory stage of a production system that generally relate to production of the blanks or work parts for the next machining stage. Metal-forming technologies imply the formation of blanks or work parts through casting or plastic deformation of the raw material and refer to a group of manufacturing methods. The forming processes do not lead to a significant change in mass, composition or volume of the material by which the given shape of the work part is converted to another shape. Metal-forming machinery by plastic deformation is represented by a variety of types of machines. Typical machinery of cold- and hot-forming technology are forging machine, hammer, press and bending machine, press-forming machine, shearing machine, blanking and punching machines,

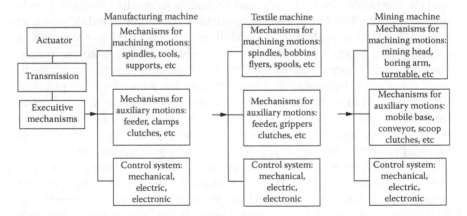

Figure 1.2 Structure of industrial machines.

rolling machine which is defined as pressurised metal forming, using one or more revolving tools, etc. Metal-forming processes produce work parts or blanks with low accuracy and generally cannot provide the desired quality that fit for assembly processes of mechanisms and machines. Following advantageous characteristics of a forming technology are high productivity and short production time and good mechanical properties of the component.

Metal-removing technologies. Machining is an essential process throughout engineering design and manufacturing industries. Machining is the most important one among manufacturing processes. Machining is the process of finishing machine parts with the desired dimensions and surface roughness by gradually removing the excess material from the preformed blank with the help of cutting tools. All machine parts in engineering, such as housings, shafts, gears, bearings, clutches and screws, need dimensional and form accuracy and good surface finish for serving their purposes. Final finish operations for machine parts are performed by grinding processes. Machining to high accuracy and finish essentially enables a product to fulfil its functional requirements, improve its performance and prolong its service. To remove excess metal during the machining process, shaping parts are used for different types and designs of cutting tools. The most important traditional machining processes are turning, boring, milling, drilling, reaming, threading, shaping/planing, broaching and grinding that produce geometrical surfaces like flat, cylindrical or any contour on the preformed blanks. Except known one, there are many new recently developed technologies such as electrical discharge machining, electrochemical machining, electron beam machining, photochemical machining, ultrasonic machining, plasma cutting and water jet cutting, rapid prototyping machining and others.

Each machining technology is a basis for the design of the machine tools that are key to industrialisation of a country. The value of machine tools is in the areas of rigidity, accuracy and precision, efficiency, productivity and ability to produce parts ranging from simple to complex pieces and different shapes and sizes. The modern manufacturing machines and systems are represented by numerous designs as single machine tools, multi-station technological modules, automated lines, etc. for implementation of manufacturing processes. The structures of manufacturing machines and systems are also having big variances with parallel, serial and mixed workstations or machine tools arranged according to the technological process for machining products. However, the complex structures of manufacturing systems conflict with their reliability problems. These manufacturing systems have some degree of reliability for mechanical, electrical and electronic units and machines. Any failure of such components leads to downtime of expensive production systems and reduces their productivity. This problem presents the challenge for engineering that should be solved for the sake of industrial progress.

Analysis of industrial machines demonstrates that numerous variants of machinery designs and nevertheless detailed consideration of them manifest similarities between most of the designs. Evolution of machine tools over the past centuries results today in the form of machine tools with CNC, machining centres of multi-axes control that combine turning, milling, grinding, drilling, etc., and material handling into one highly automated manufacturing machine. CNC machine tools can produce complex machine parts of various sizes and shapes with high levels of precision. Other result of machine evolution is an automated production line that represents a system with complex design combining several machine tools or workstations, transport and buffers system and computerised control. Automated production lines are based on segmentation and balancing of the manufacturing process that arranged the sequence individual processing tasks at the workstations so that the total time required at each workstation is approximately the same. The segmentation of manufacturing process on defined number of serial workstations enables to dramatically enhance productivity rate. It is a way to mass produce goods quickly and efficiently for all of the manufacturing processes. Such revolutionary solution enables reducing total labour hours for finished product and reduces the unit cost for manufactured goods in large quantities.

Automation. Modern tendency of manufacturing processes is automation of machines and systems. Automation technology has revolutionised manufacturing processes since the Industrial Revolution, where they are used for machining and other processes based on principles of mechanical automation. The development of electrical and electronic industries enabled using the automation technology more intensively and effectively especially by computer-related technologies. Automation technology is a natural result of engineering evolution, and today automated manufacturing process is represented as automated factories with robotics. However, the level of automation of production processes at manufacturing sectors is different. There are high and poor automated sectors in production systems. Regular more or less developed production systems have three main sequence stages: preparatory, machining and assembly stages. Analysis of production systems for different companies by criterion of automation is depicted in Figure 1.3. Preparatory stage that consists of the production processes of work parts by casting, pressing, forging, cutting operations, etc., is represented in average 10%–15% automated machinery. Machining stage of work parts by turning, boring, drilling, grinding operations, etc., consists in average 75%–80% automated machine tools and automated lines. Assembly stage has the lowest level of automation, i.e. in average 5%–10% automated workstations. Similar percentage of machinery is related to production stages.

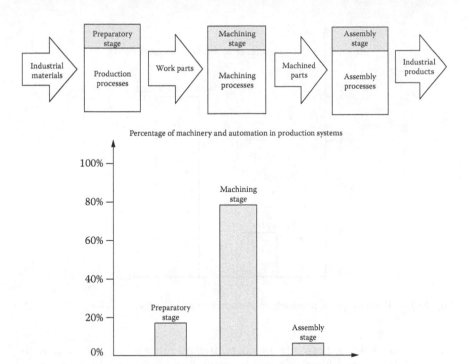

Figure 1.3 Production stages and percentage of machinery and automation.

The percentage of a machinery at production systems depends on their types, i.e. mass, batch and workshop. Industrial statistics demonstrates the following picture of the production system presented in Figure 1.3.

The picture about the percentage of machineries at the types of production systems is presented in average that have differences in types of industries, i.e. metal forming, textile, transport, etc. Figure 1.4 demonstrates that batch and job-shop production systems are most spread in industries compared with mass production systems. Hence, the type of machinery is spread accordingly. At the batch and job-shop production systems, most machineries are manually controlled. The automatic machines and automated lines are the machineries mostly used at mass production systems. These two types of machineries possess different properties and indices, especially in productivity rate. Production systems of high productivity are made conditional by automation of processes.

Advantages commonly attributed to automation include higher productivity rates, more efficient use of materials resulting in less scrap, better product quality, improved safety, shorter workweeks for labour and reduced factory lead times. Higher output and increased productivity have been two of the biggest reasons in justifying the use of automation.

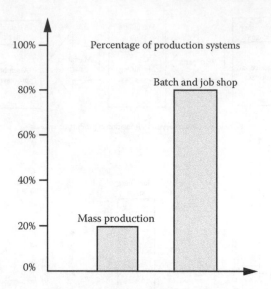

Figure 1.4   Percentage of production systems in industries.

Automated systems typically perform the manufacturing process with less variability, resulting in greater control and consistency of product quality.

Simple machining motions that are mainly linear and rotary and a combination of them represent the reason of the simplicity of automation for machining process. Automation of processes in preparatory and assembly stages is not a simple problem due to complexity of processing motions. Automation of machining processes expressed in productivity increase. Hence, the preparatory and assembly stages of manufacturing process should also increase the productivity rates that should be equal to the productivity rate of a machining stage. This situation with automation of machining stage leading to an increase in the number of employees at the first and last stages can result in an increase in the number of employees for the companies. The known statement in industries that automation of manufacturing processes lead to decrease in the employees is true, but it should be pointed that this decrease is local and mainly related for machining stage. The modern computerised engineering enables automation of any complex processes, but their high cost and related properties represent restrictions and is a subject of economic efficiency for companies.

Joining technologies. Manufacturing processes like welding, brazing, soldering, adhesive and acoustic bonding, clinching, screwing, riveting and bolting are based on joining technologies. Every manufactured product contains joints, either to assemble similar materials into a more complex shape or to attach dissimilar materials to create a composite product incorporating different material properties. To apply which joining method for manufacturing process is appropriate depends on its advantages and disadvantages.

Manufacturers produce different types of joining machines that can be applied to any application requiring the joining of metals and materials.

Assembling and packaging processes. Final technologies in manufacturing processes of products are assembling and packaging processes, which are the most complicated stages of a production system in terms of data flow. The complexity is due to interfacing with a large number of components to be assembled. Assembling processes require access to geometry, connectivity, material and fabrication data with balancing of the assembly process technology. Most of the assembly technologies in industries are implemented manually with minor mechanisation and automation processes. The balancing of assembly process is to optimise planned assembly lines, minimise costs and maximise productivity rate and efficiency. Assembly process is organised such that the motion of each worker is minimised to perform one simple operation at a time required at each station. This allows the basic aspects of the assembly process to be implemented without excessive delay of industrial machines and tools, work of employees for continuous flow of machine parts in mass-production processes

1.3   Reliability attributes for industrial machines and systems

Today's modern production systems have become increasingly complex to design and build, while the demand for reliability development continues. Reliability is one of the most important attributes in all production systems and plays a key role in cost-effectiveness of industrial systems. Growing international competition has increased the need for all designers, managers, practitioners, scientists and engineers to ensure a level of reliability of their product before release at the lowest cost. The interest in reliability has been growing since the Second World War, and this trend will continue during the next decade and beyond.

Intensification of industrial processes magnifies their reliability problems. The efficiency of expensive production systems, such as automated lines with complex design, largely depends on the reliability of its main mechanisms and units. Decrease in the reliability of complex automated manufacturing systems results in a dramatic decrease in productivity. This reliability problem is not a new one for the industry and its engineering attributes are well known, making it possible to describe manufacturing problems analytically. The theory of reliability provides standard attributes and modes of calculation, which can be used for mathematical modelling of reliability of manufacturing machines. However, known attributes and equations of reliability describe properties of an industrial machine separately from its productivity rate, economic efficiency and other indices. Several publications are dedicated to studying the productivity and reliability attributes of automated lines with different designs. Analysis of these publications and equations for the productivity of machines with complex designs demonstrates that there are simplifications and assumptions that do not capture the full meaning of some components. This simplified approach cannot give exact mathematical models of machine productivity; therefore, there will always be differences between theoretical and actual productivity.

Reliability engineering is a field that deals with the study, evaluation and life cycle management of machines and systems. The importance of reliability engineering for industries is underlined by the standard definition. Reliability is the ability of a system or component to perform its required functions under stated conditions for a specified period. Reliability engineering is described by several attributes in terms of the probability of failure, the frequency of failures, or in terms of availability and maintainability. Reliability is associated with unexpected failures of products or services and understanding why these failures occur is a key to improve reliability. It is therefore a measure of engineering problems and to quantify reliability involves the use of statistics and more specifically probability theory.

Engineering systems are designed to operate well in the face of uncertainty of characteristics of components and operating conditions. Manufacturing processes are subject to unpredictable variations based on the probabilistic approach that represents random events in mathematical models of uncertainty. The reliability and probability theories are put into practice for computing availability of machinery. This is the reason that engineers need to know and use the rules, theorems and axioms of reliability and probability theories for solving complex industrial problems.

The growth in design complexity of machinery in most industries leads to an intensification of the failure rates that results in a decrease of the productivity rate, and as a consequence, the production process has become less effective. The low reliability makes the following aspect more important: prediction of the reliability level of machinery and its major parts and prediction of the availability of machinery. These predictions can only result from careful consideration of reliability and maintainability factors at the design stage of complex industrial engineering. The are many reasons why failures occur, which include different sources of different nature like insufficient design margins, incorrect use industrial environments, undetectable defects, human error and abuse, unavoidable failures, inadequate or improper preventive maintenance, limited life components, wear out due to aging and friction, misalignments, corrosion and creep, incorrect overhaul practices, etc.

To ensure good reliability, the causes of failure need to be identified and eliminated. Hence, the objectives of reliability engineering are to prevent or reduce the likelihood or frequency of failures, to identify and correct the causes of failures, to determine ways to reduce failures, to apply methods of estimating the likely reliability of new designs, and to analyse reliability data.

The failure probability for the generality of industrial system components is estimated by the failure probability profile, work time and repair-time distribution of the constituent components. This is an essential requirement for the prediction of engineering system reliability and estimation of availability, maintainability, and the level and cost of corrective and preventive maintenance. The following steps perform evaluation and prediction the reliability of complex engineering system:

- Engineering system with units and items is ranked according to their function.
- Determination of the connection level for the system's units and items.
- Analysis and calculation of the reliability attributes for all components and levels and definition of their functional configurations in the system.
- The result of reliability of the engineering system is used in its improvement, in the calculation of reliability attributes and in the prediction of maintenance workload.

Reliability engineering for complex systems requires a different, more elaborate systems approach than for non-complex systems. Reliability engineering may involve functional failure analysis, proper and useful requirements specification and analysis, hardware and software design, manufacturing, testing, maintenance, transport, storage, spare parts, operations research, human factors, technical documentation, data and information acquisition/organisation, etc. Effective reliability engineering requires understanding the basis of failure mechanisms for which experience, broad engineering skills and good knowledge from many different special fields of engineering, such as tribology, stress, fatigue, thermal, electrical, chemical and design.

Many analytical techniques are used in reliability engineering, such as reliability hazard analysis, failure mode and effects analysis, fault tree analysis, material stress and wear calculations, fatigue and creep analysis, finite element analysis, reliability prediction, thermal stress analysis, corrosion analysis, human error analysis, reliability testing, statistical uncertainty estimations, Monte Carlo simulations, design of experiments, reliability-centred maintenance, failure reporting and corrective actions management. Because of the large number of reliability techniques, their expense and the varying degrees of reliability required for different situations, most projects develop a reliability program plan to specify the reliability tasks that will be performed for that specific system. The most common reliability program tasks are documented in reliability program standards. Failure reporting analysis and corrective action systems is a common approach for product/process reliability monitoring.

Reliability engineering focuses on productivity decrease caused by system downtime, repair equipment, personnel and cost of warranty claims, which may cause production loss. High reliability levels are also the result of good engineering, attention to detail and almost never the result of reactive failure management. However, high reliability of a machine is subject to its high cost, such as high accuracy of machining, which has an impact on the economic efficiency of engineering systems. This statement leads to develop the system for reliability tolerances that should be optimised by criterion of economical attributes.

1.3.1 Measuring and modelling the system reliability

Reliability theory presents several attributes for measuring reliability indices of machines and systems. The productivity theory of industrial engineering considered some rules, regulation and components of reliability and probability theories that have right application to manufacturing systems with complex structures. These components are presented next.

The probability density function of random events is calculated as follows:

$$p(x_1 < x < x_2) = \int_{x_1}^{x_2} f(x)dx \tag{1.2}$$

where $f(x)$ is the probability density function of a random event x.

The cumulative distribution function, $F(x)$, gives the probability that a measured value of random events will fall between $-\infty$ and x:

$$F(x) = \int_{-\infty}^{t} f(x)dx \tag{1.3}$$

The cumulative reliability function, $R(x)$, i.e. there is no failure in the interval (0 to x) gives the survival function:

$$R(x) = 1 - F(x) = 1 - \int_{-\infty}^{t} f(x)dt \tag{1.4}$$

There are three continuous distributions of events in reliability engineering: the exponential, Weibull and lognormal distributions. The exponential distribution is used most commonly and practically in reliability engineering. The Weibull distribution is used in case by which some percentage of the population can be expected to fail. The lognormal distribution is more versatile than the normal distribution and is a better fit to reliability data, such as for populations with wear-out characteristics.

In manufacturing engineering, mechanical systems designed with components that have hard connections in series, parallel and mixed modes. Multiple components make up systems that should be evaluated by the reliability level of a system. The components are connected hard together determines the system reliability model that should be used. In mechanical systems of complex structure, the failure of one component leads to a stop of all systems. For mechanical systems with independent components for which the failure behaviour in the system is quite uninfluenced by that of the others, the survival probability of the system $P_s(t)$ at time t is given by the product of the separate survival probabilities of the components at the time t and is presented by the following equation:

$$P_s(t) = P_1(t) \times P_2(t) \times P_3(t) \times \cdots \times P_n(t) \tag{1.5}$$

where $P_n(t)$ is the failure probability of a component n.

In addition, the system reliability as the survival function is presented by the following equation:

$$R_s = R_1 \times R_2 \times R_3 \times \cdots \times R_n \tag{1.6}$$

If the times to failure of the components behave according to the exponential probability density function, then the overall probability density function of the times to failure is also exponential and presented by the following equation:

$$F(t) = (\lambda_1 + \lambda_2 + \lambda_3 + \cdots + \lambda_n) \times \exp\left[-(\lambda_1 + \lambda_2 + \lambda_3 + \cdots + \lambda_n)t\right] \tag{1.7}$$

where λ_n is the failure rate of n components.

The system reliability is computed by the following equation:

$$R_s = e^{-\lambda_1 t} \times e^{-\lambda_2 t} \times e^{-\lambda_3 t} \times \cdots \times e^{-\lambda_n t} \tag{1.8}$$

The failure rate of the system is computed by adding the failure rates together:

$$\lambda_s = \lambda_1 + \lambda_2 + \lambda_3 + \cdots + \lambda_n = \sum_{i=1}^{n} \lambda_i \tag{1.9}$$

Practically, standard reliability attributes are maintainability, availability and failure rate or mean time between failures, which is the inverse of failure rate. Maintainability is the ability of an item, under stated conditions of use, to be retained in, or restored to, a state in which it can perform its required functions, when maintenance is performed under stated conditions and using prescribed procedures and resources. Maintainability is expressed as the mean time to repair, m_r.

The mean time to failure (MTTF) of a system is presented by the following equation:

$$\text{MTTF}_s = 1/\sum \lambda_i \tag{1.10}$$

Failure rate is expressed as the mean time between failures, which is the inverse of failure rate:

$$\lambda = 1/m_w \tag{1.11}$$

where m_w is the mean time between failures.

Availability is the probability that a system is available for use at a given time, which is a function of reliability and maintainability. It is operating time divided by load time, which is the available time per period of observation minus the planned downtime. Failure is the termination of the ability of an item to perform its required function. The known attributes of reliability are classified into two groups: single attributes that characterise one aspect of machine failure and attributes that include several single attributes of reliability, one of which is availability. The availability of a system is represented by the following equation:

$$A = \frac{\theta_w}{\theta_w + \theta_i} = \frac{m_w}{m_w + m_r} = \frac{1}{1 + m_r \lambda} \tag{1.12}$$

where A is the availability of a machine, θ_w is the time of a machine work and θ_i is the idle time when a machine does not work including repair and maintenance time, m_w is the mean time between failures, m_r is the mean repair time and λ is the machine failure rate.

These attributes of reliability are used to develop analytical equations for the productivity rate of machines and systems with simple and complex designs like automated lines. The known equations of productivity rate for different designs of automated lines include technological parameters (machining time), technical parameters (auxiliary and idle time, capacity of buffers, etc.) and structural parameters (number of serial, parallel and mixed workstations and number of sections of automated lines). Equations contain a reliability component that is machine idle time, which can be transformed into standard attributes of reliability.

Industrial machines with complex designs should be calculated by holistic analytical approach that combines mathematical modelling of productivity and reliability. Such modelling makes it possible analytically to describe manufacturing problems using reliability attributes and thereby enhance output. However, known attributes and equations of reliability describe properties of a machine separately from its productivity rate. The theory of reliability is based on several principles derived from electrical and electronic systems using serial, parallel and combined models of components. The theory of reliability does not consider the system component as a productive unit. These rules of the reliability theory should be well thought out before use for industrial machines with complex designs, whose operational principles are quite different.

The productivity of machines and systems has a direct relationship with the reliability of their units, which can be described analytically. Analysis and mathematical modelling of machine productivity makes it possible to embed standard attributes of reliability as components of productivity rate equations for different designs and structures of industrial machines. Analytical study of the main attribute of reliability (availability) in machines with complex designs demonstrates that, except standard attributes of machine units' reliability, the structural and technical parameters of machines are the components of availability. This important information can present the availability of a manufacturing system in the new capacity. The productivity and efficiency of manufacturing systems depend on availability, which includes structural information of complex systems.

The tendency of engineering progress is represented in the form of producing industrial machines with more and more complex designs, whose efficiency must be evaluated by the criterion of productivity. Otherwise, the efficiency of industrial machines that is calculated based on stress, kinematics, dynamics and accuracy is deemed useless. Clearly, industrial machines of low productivity are not effective and are therefore not profitable for companies. Manufacturers prefer to use machines of high productivity that fabricate qualitative products. Analysis and synthesis of engineering solutions based mainly on the criteria of productivity rate, quality of products, cost and system flexibility express all parameters of manufacturing processes.

Mathematical models for industrial machines' productivity are different and depend on several technical and technological data, and are used according to the level of consideration of the manufacturing processes. The mathematical models of productivity consider the technological and technical aspects with reliability attributes of machines with random failures. The aspects of management or maintenance of machines at prescribed planned overhaul repair time is considered separately. The management factor considers the quality of the production process functioning that does not depend on the productivity rate of a machine or its reliability with random failures. The concept of maintenance considers the machine when it is out of work, i.e. the planned repair and service of machines that are stopped for overhaul processes, which does not include the random stops of machines. Neither the concept of management nor maintenance takes into consideration the reliability attributes of a machine. However, consideration of a machine's lifespan is included in the general attribute of machine efficiency.

Preliminary analysis of mathematical models for the productivity of automated lines shows that the reliability component of models can be presented by standard attributes (failure rate, mean time to failure and mean time to repair). These reliability attributes as well as technical and structural parameters of different types of automated lines can be analytically combined and presented in the form of availability in mathematical models of productivity rate.

Hence, the availability of industrial machines is an integrated attribute presented by mathematical models that include single attributes of machine units' reliability, and the structural and technical parameters of complex production systems. Productivity theory for industrial engineering presents mathematical models of productivity rate with the availability of single machines and complex automated lines. Mathematical models of productivity rate demonstrate the interdependency of productivity, reliability, technological and technical parameters and the structure of manufacturing systems.

Bibliography

1. Altintas, Y. 2012. *Manufacturing Automation*. 2nd ed. Cambridge University Press. London.
2. Benhabib, B. 2005. *Manufacturing: Design, Production, Automation, and Integration*. 1st ed. Taylor & Francis, Markel Dekker Inc. New York, Basel.
3. Birolini, A. 2007. *Reliability Engineering: Theory and Practice*. 5th ed. Springer, New York.
4. Boothroyd, G. 2005. *Assembly Automation and Product Design*. 2nd ed. Taylor & Francis E-Library.
5. Elsayed, A.E. 2012. *Reliability Engineering*. 2nd ed. John Willey & Sons. Hoboken, NJ.
6. Groover, M.P. 2013. *Fundamentals of Modern Manufacturing: Materials, Processes, and Systems*. 5th ed., (Lehigh University). John Wiley & Sons. Hoboken, NJ.
7. Hansen, R.C. 2005. *Overall Equipment Effectiveness*. Verlag, Industrial Press Inc. New York.
8. Jorgensonn, D.W. 2009. *The Economics of Productivity*. Edward Elgar. Northampton, MA.
9. Kalpakjian, S., and Schmid, S.R. 2013. *Manufacturing Engineering & Technology*. 7th ed. Pearson. Cambridge.

10. Mehta, B.R., and Reddy, J.Y. 2014. *Industrial Process Automation Systems.* 1st ed. Butterworth-Heinemann, Elsevier. New York.
11. O'Connor, P.D.T., and Kleyner, A. 2012. *Practical Reliability Engineering.* 5th ed. John Wiley & Sons. West Sussex.
12. Ortiz, C. 2006. *Kaizen Assembly: Designing, Constructing, and Managing a Lean Assembly Line.* Taylor & Francis E-Library.
13. Rekiek, B., and Delchambre, A. 2006. *Assembly Line Design.* Springer. London.
14. Rao, R.V. 2013. *Manufacturing Technology – Vols. 1; 2.* 3rd ed. McGraw-Hill Education. New York.
15. Schey, J. 2012. *Introduction to Manufacturing Processes.* 3rd ed. McGraw-Hill Education. New York.
16. Shaumian, G.A. 1973. *Complex Automation of Production Processes.* Mashinostroenie. Moscow.
17. Volchkevich, L. 2005. *Automation of Production Processes.* Mashinostroenie. Moscow.

chapter two

Conceptual principles of productivity theory for industrial engineering

History of engineering demonstrates that the moving force for engineering evolution is a constant requirement for sustainable improvement; increasing productivity rate and quality of products, services and efficiency of processes. Engineers of the past century created several brilliant methods of productivity increase by uniform decomposition of complex technological process on several serial components and implementation of machine tools. Contemporary manufacture of machines and systems demonstrate that attempts to increase the productivity rate of manufacturing systems are encountered on several technical and technological restrictions that should be resolved analytically. Productivity theory for industrial engineering collected all achievements in analytical and practical experience in formulation as the productivity rate for different designs of manufacturing systems. This theory describes laws and reasonable dependencies in the design of machinery, demonstrates relationships between productivity, reliability, technological and technical parameters and the structure of serial and parallel–serial arrangements of machines and systems. Increase in the productivity rate can be achieved by intensification of the machining regimes for multi-tooling processes, but it also leads to an increase in the failure rate of machine tool units. The mathematical model for the optimal cutting speeds with the maximal productivity rate solves these controversial results. The principle theory of industrial productivity with holistic mathematical models enables defining maximal productivity rate, optimal machining regimes and structures for any industrial machines and systems.

2.1   Introduction

Manufacturing processes in industry are complex and involve activities from different departments that include engineering design, technology, management, marketing, economics, etc. Industrial processes are evaluated using many indices, with the two primary indices being the productivity of manufacturing systems and the quality of products. Of course, the economic efficiency is the final index for taking the decision for launch of a production process. The work of different departments is concerted and coordinated to carry out the main indices of the production system. This complex coordination has causative, consequential and effective links that should be formulated using mathematical models. Such models are useful in evaluating the efficiency of production system components, and numerically, they demonstrate the level of their influence on factory outputs. Mathematical models for the productivity of industrial machines are important because they facilitate the evaluation of a manufacturing system based on efficiency. Otherwise, the activity of different production departments and systems is deemed useless. In manufacturing engineering, progress is often evaluated in the form of mathematical models of productivity systems, whose output should be predicted with high accuracy.

In the engineering field, mathematic theory and application are basic tools for design and analysis of industrial processes and manufacturing engineering as well with mathematical model or mathematical analysis. Mathematical method is generally combined with statistical method in analysis stage, which is graphical, or chart presented with the support of mathematical calculation. This method considers mathematical and statistical analysis method, since it involves the mathematical model application and statistical analysis method. Because of the high complexity of industrial processes, especially the manufacture of automated lines with complex structure, the demand for a less complicated, convenient and effective engineering mathematical and statistical analysis method is strongly needed for analysis and sustainable improvement of the productivity rate of manufacturing systems.

During the centuries, industries accumulated experience in creating different manufacturing machines and systems for production of various goods, parts and products. All these diversity of machines and systems with different designs are solving the pivoted problem to increase the productivity rate with high quality of output products. Productivity theory for industrial engineering collected all achievements in analytical and practical experience formulated the productivity rate for different manufacturing systems that has simple or complex designs. Productivity theory for industrial machines and systems discovers laws and reasonable dependencies for design of new manufacturing machinery. Mathematical models of productivity rate for machines and systems demonstrate relationships between productivity, reliability, technological and technical parameters, machining regimes and the structure of machines and systems.

The modern industrial machines and systems are considered as systems that contain different mechanisms and constructions based on mechanical, electrical and electronic units. The structures of manufacturing machines and systems also have a big variance with serial and parallel units stations arranged according to the technological process of machining products. These manufacturing systems have different indices of reliability units with variable physical principles of work. Any failure of such components leads to downtime of an expensive automated line of complex structure.

The main goal of theory of industrial productivity is to design manufacturing machines and systems with high output and high quality of products. Conceptual principles of productivity theory for industrial engineering are the following [14,17]:

- Analysis of all factors based on the technology, design, structure, reliability and exploitation of manufacturing machines and systems that influence on productivity rate.
- Production time is the primary component of production systems that should be considered by three aspects and described by mathematical models:
  a. Processing time that spends on machining, assembly or transformation of the objet in space by industrial machines.
  b. Auxiliary time that spends on load and remove of a work part, clamp and rerelease in machining area that is part of a production time.
  c. Idle time of the machines and systems due to their reliability, managerial and organisational problems.
- Increasing the productivity rate of manufacturing processes is implemented by intensification, segmentation and balancing, and duplicating of technological processes that are limited by criterions of quality of fabricated products and structural design of manufacturing machines and systems.
- Manufacturing systems are maintained at workable condition by the team of technicians and minimal idle times of machines do not exceed the accepted tolerance.

For production process, the auxiliary and idle time of machines and systems is loss time, i.e. machine or system does not produce work parts or goods. Naturally, the non-productive times should be decreased by engineering and managerial activity improves the efficiency of a factory. The principles described earlier enable developing the mathematical models for productivity rate, optimisation of machining regimes and structural designs of industrial machines and systems and directions to increase productivity rate. The productivity theory formulated and combined the diversity of mathematical models for the productivity rate of manufacturing machines and systems of different design based on a variety of technological processes and their parameters.

Different industries used different technologies, machinery and implemented different works. Hence, the mathematical models for productivity machines and systems for these industries will have specific parameters that reflect their particularities in technologies. If manufacturing and other similar industries are implementing repeatable work, it represents the machining or fabricating products at prescribed time, whereas in transport industries, there are fulfilled repeatable works to deliver objects to prescribed place and time by airplanes, trains, buses, cars, etc. Of course, other industries have their own specific properties of technological processes that should be included in mathematical models of the productivity rates. For example, the city bus motions from one bus stop to the other one represent the repeatable work of the bus traffic system that belongs to the transport industry. Mathematical models for productivity rate of the city bus system should be derived for its rush hours and regular one. Each industry has its own specific technologies and peculiarities. Nevertheless, it is easy to apply the productivity theory for industrial engineering for any type of industries and processes, i.e. for manufacturing, mining, transport, food, chemical, agricultural, etc. industries. This main property of productivity theory for industrial engineering expresses its universality, which can be automated with exact timing processes or manually controlled systems, where the timing components of production processes are calculated on statistically average data.

Considering the processes of industries development in evolution, it is easy to trace that each historical step of engineering evolution is solving the problems of increase in the productivity rate and quality of products, services, etc. However, this evolutionary process of engineering linked with other problems. Increase in productivity rate of machinery leads to change in the numbers of employees, increasing the cost and complexity of engineering, etc. In our modern life, decrease of numbers of employees in real automated industries is not a problem, because there are many branches of economics with manual human activity that need mechanisation and automation to enhance productivity rate. History of engineering demonstrates that the moving force for engineering evolution is a constant requirement in sustainable improvement, increasing productivity rate and quality of products, services and efficiency of processes. These criteria are decisive for evaluation of progressiveness and perspectives of new engineering.

2.2 Technological processes and balancing are the basis for structural designs of manufacturing machines and systems

In engineering, all mechanisms, machines and systems are assembled from the components, and machine parts with different designs can have simple and complex constructions (Figure 2.1). For machining, these parts created different technological processes that implemented on the

Figure 2.1 Typical machine parts.

machine tools and manufacturing systems with different structures. Simple technological processes are conducted on one machine tool with simple designs. Complex technological processes for machining the parts cannot be conducted at one machine tool due to its technical restrictions. The parts with complex designs are machined on the manufacturing systems of different structures that can contain multiple numbers of machine tools and workstations.

Each type of technological processes has its own properties described by machining regimes that is applied for manufacturing products. Machining regimes for metal-cutting technologies are mainly presented by the cutting speed, feed rate and depth of cut. The willingness to get high productivity of machine tools leads to the intensification of machining regimes. However, this way has several restrictions that should be considered in detail. Increase in the feed rate and depth of cut in machining process decreases the quality of a machined work part. Increasing the cutting speed enables to increase the productivity rate, but decrease the tool life, i.e. increases the numbers of tool replacement and failure rate of other units. These properties of metal-cutting processes demonstrate that there are some optimal machining regimes, which can give maximal productivity rate and necessary quality of the work parts. Mathematical models of productivity theory for the industrial machines and systems can solve optimisation of the machining processes by criterion of maximal productivity rate.

The increase of productivity rate is always dominating in industries that have simple and complex manufacturing processes. Engineers of the past century created brilliant methods of productivity increase by uniform decomposition of complex technological process on several serial short components and distributing its implementation on serial sequence machine tools or workstations. This method dramatically improved the productivity of manufacturing systems. However, at the same time, such production processes are implemented by increasing the number of employees, machines, consumption of energy, etc. New method created new problems of structural design of manufacturing systems that should be solved by mathematical methods of multi-parametrical analysis of variables for production processes. The method of uniform decomposition of the technological process with relatively equal machining time on sequence of machine tools or workstations is called balancing. This method is used in all of the manufacturing processes. The decomposition of technological process enables enhancement of the productivity rate of the manufacturing systems quickly and efficiently, because each machine tool or workstation implements a simple component of the complex technological process. The work parts that are machined on the manufacturing systems are passing consequently all workstations and getting full technological processes. Each workstation of simple design is conducted as a simple component of technological process for a short time. The total time of technological process is balanced as much as possible uniformly on the workstations. However, practice demonstrates that any manufacturing process always has the serial bottleneck workstation with longest cycle time that defines the productivity rate of the production system. Other peculiarity of balancing is that the decomposition of technological processes cannot be conducted endless due to several restrictions. The machining of a high-quality surface should be conducted by a single cutter. Multi-tooling process of such surface leads to increase in the surface roughness and decrease in the accuracy of machining. This circumstance restricts decomposing of machining process and limits the productivity rate. Additionally, multi-tooling process leads to increase in the cost of machine tools or workstations.

Production experience in manufacturing industries demonstrates that different machined work parts can be manufactured by having simple and complex technological processes. The complex and long technological process for machining the work parts is decomposed on definite number of workstations or machine tools that depend on the required productivity rate for the manufacturing system. The simple and short technological processes that cannot be decomposed, like stamping, coining, forging or others, but demand to increase productivity rate leads to parallel arrangement of such machine tools and workstations. Hence, workstations or machine tools can be arranged in the parallel, linear and mixed (parallel–linear) structures of manufacturing systems that depend on a necessary value of productivity rate and on the complexity of technological processes.

Successful manufacturing system designs must be capable of satisfying the strategic objectives of production system. Main objectives of manufacturing systems design is high productivity of machinery and high quality of manufactured products.

Manufacturing systems design is the process of defining the structure, components, modules, interfaces and data for a system that satisfy specified requirements for productivity. Systems design could be seen as the application of systems theory to product development. The manufacturing system design relates to the actual output processes of the system. This is laid down in terms of how data is input into a system, how it is verified, how it is processed and how it is displayed as manufacturing system design. The following are the requirements about how the system is decided:

  a. Necessary productivity rate.
  b. Appropriate technology.
  c. Storage of products.
  d. System control and backup or recovery.

Productivity rate requirement is solved by intensification, segmentation and duplication of technological process. Processing requirements enable defining appropriate technological process, machine tools, workstations, transport system and structural design of production systems. Solving these problems simultaneously results in achievement of minimal cost and high quality of the product. Storage requirements are solving problems of machined parts store between machine tools, automated lines and workshops. System control and backup or recovery enables to keep the manufacturing system in workable conditions. Making these decisions in a way that supports high-level objectives of a manufacturing system requires an understanding of how design issues affect the interactions among various components of a manufacturing system.

Manufacturing systems and their structures are categorised into several groups regarding to work of machine tools and workstations. First system is represented by machine tools that work independently from each other. Automated lines with workstations that have some dependency from each other and transport and control system represent the second system. Other manufacturing systems are represented by a combination of independent and dependent work of automated lines with different structures.

Structural designs of manufacturing systems are represented by parallel, serial and parallel–serial arrangements. In turn, all arrangements are designed into linear or circular constructions. This diversity in arrangements and designs is represented in Figure 2.2.

Automated manufacturing systems are represented by linear structures with straight flow conveyor or other transport system for all types of arrangements of manufacturing systems, while circular structures are different for parallel and parallel–serial and serial designs of automated liners. Automated lines of parallel and parallel–serial structures are designed into rotor-type system with moving workstations on a rotor machine. Automated lines of serial structures and circular design arranged with the rotor table of indexed turn implement transport function for work parts. These peculiarities in design of the manufacturing systems represented mathematical models for productivity rates as well as peculiarities in technologies and technical parameters.

Engineering practice in different industries developed several solutions for linear and circular designs of manufacturing systems. Manufacturing industries have short and heavy machining processes, while the production systems have a circular design with small numbers of workstations or machine tools. For electric, electronic, textile and other

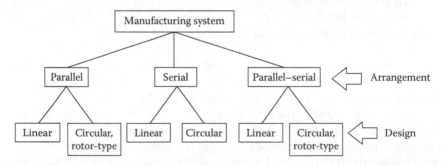

Figure 2.2 Diversity in arrangements and designs of manufacturing systems.

industries with light manufacturing processes, the number of workstations in the circular lines can be bigger.

The linear structure of manufacturing systems is preferable for complex technological processes implemented on the multi-station or multi-machine tool systems. The logical design of a system pertains to a graphical representation of the structure of the manufacturing system. This is often conducted via modelling, using a graphical model of the actual system. For theoretical analysis, the productivity rate for manufacturing systems is necessary to present the machine tools, multi-station manufacturing systems, production and automated lines, etc. by simple symbolic pictures. This type of symbolic presentation of machinery enables conducting analytical approach and derives mathematical models of productivity rate for manufacturing systems with different designs and structures. These systems can be arranged using the following structures:

    a. System with independent machine tools that implement one technological process and arranged in parallel structure (flow line of parallel structure with independent machine tools).
    b. System with parallel machine tools (workstations) combined in one system that implement one technological process by all workstations and arranged in parallel structure (automated line of parallel structure).
    c. System with independent machine tools that implement different sequence components of technological process and arranged in serial structure (flow line of serial structure with independent machine tools).
    d. System with machine tools (workstations) combined in one system that implement components of sequence technological process and arranged in serial structure (automated line of serial structure).
    e. System that presented parallel and serial flow lines with independent work of the machine tools (flow line of parallel–serial structure).
    f. System that presented parallel and serial automated lines (automated line of parallel–serial structure).

Schematic and symbolic pictures for all types of structural designs include the single independent machine tools, and different structures of manufacturing and automated lines are presented in Figure 2.3a–f.

The structures provided earlier of manufacturing systems have different properties in a balancing of technological processes, productivity rate, reliability, area of application, number of employees engaged, service system, cost, etc. The main index of the efficiency of a manufacturing system is productivity rate, which depends on the many factors that include the technological, constructional, managerial and reliability components, the latter one is expressed in downtime of machines and workstations. At manufacturing line with independent work of workstations or machine tools, a random downtime of any machine tool will not lead to stops of other machines. Independent machine tools work and have downtimes, and machined products are stored in space between two sequence neighbouring machine tools. Independent machines consume work parts stored by other machine tools. At the automated line of different designs, downtime of any workstation or mechanism leads to stop the whole line and reduces productivity rate. Which design and structure of manufacturing systems is preferable and optimal represents the complex problem that can be solved by the mathematical methods of productivity theory for industrial machines and systems. Mathematical modelling of the technological, technical and constructional properties enables defining the best solution for manufacturing systems by defined criterions.

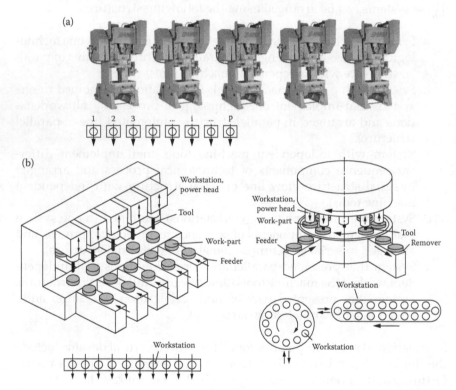

Figure 2.3 Symbolic pictures of single independent machine tools and different structures of manufacturing and automated lines. (a) Symbolic presentation of a manufacturing line of parallel arrangement with independent machine tools. (b) Symbolic presentation of an automated line of parallel structure with linear and circular designs (rotor-type) for small numbers of workstations and linear design (rotor-type) for big number of workstations, respectively.

(*Continued*)

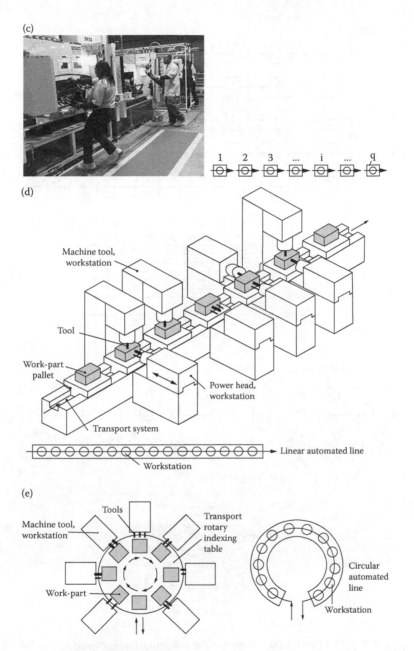

Figure 2.3 **(CONTINUED)**   (c) Symbolic presentation a manufacturing line of serial sequence arrangement with independent machine tools. (d) Symbolic presentation of an automated line of serial structure with linear and circular arrangements. (e) Symbolic presentation for a flow line of parallel–serial structure with q serial and p parallel independent machine tools.

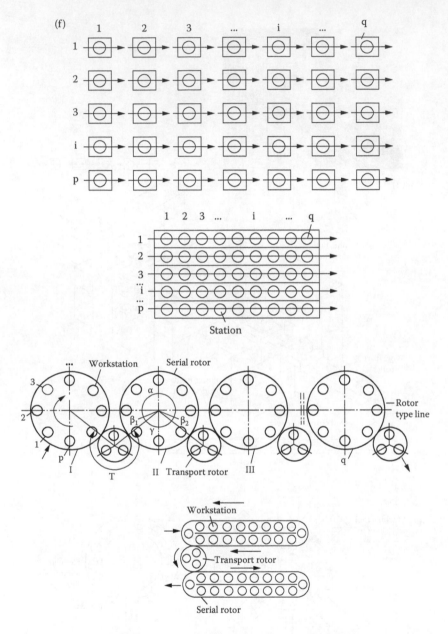

Figure 2.3 **(CONTINUED)**   (f) Symbolic presentation for production line with independent workstations and an automated line of parallel–serial structure with linear and circular (rotor-type) designs for small numbers of workstations with q serial and p parallel stations and linear design (rotor-type) for big number of workstations, respectively.

Analysis of information presented earlier enables summarising and formulating the following decisions for the sake of increasing the productivity for manufacturing machines and systems:

- New technological process that leads to increase the productivity rate for industrial system needs to design new manufacturing machines. Industrial practice demonstrates that using old design of a machine for new technology does not give high efficiency.
- Intensification of the machining process enables increasing productivity, but restriction of the physics of technological processes and quality of machining work parts limit this approach.
- Segmentation and balancing of the technological process enable increasing the productivity of the serial production and automated lines. The production line with independent machine tools occupies a large production area. The complexity in design of the automated line leads to a decrease in reliability and productivity rate. These controversial properties of the automated line are solved by analytical optimisation based on the criterion of maximal productivity rate.
- Arrangement of automated lines with parallel structure leads to an increase in the productivity of lines, but complexity of design leads to a decrease in the reliability of automated lines and limitation of the productivity rate.
- Combination of serial and parallel-automated lines dramatically increases the productivity rate, but complexity of designs leads to a decrease in the reliability of lines and productivity rate. These controversial properties of production processes lead to their optimisation by criterion of maximal productivity rate.

Presented information demonstrates that any attempt to increase the productivity rate of manufacturing systems is encountered on several technical and technological restrictions that lead to limitation of the productivity rate. These two contradicting factors are represented as engineering problem of optimisation machining processes and structural designs of manufacturing systems that is solved by mathematical modelling.

2.3   Productivity rate of a single machine tool

In industries, most of the machinery is represented by single machine tools, for which the productivity rate should be presented and described by the mathematical model. This model should reflect any changes in technology and design of a machine tool. Symbolic picture for the single machine tool is represented in Figure 2.4.

Mathematical model of productivity rate for a single machine tool should contain all technical and technological parameters of manufacturing

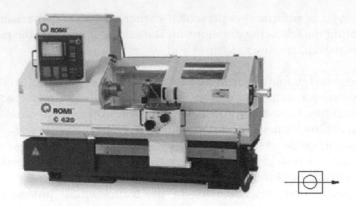

Figure 2.4 Symbolic presentation of a single machine tool.

system. This is an important fact for manufacturing industries, because the knowledge of the machine tool's parameters enables to theoretically calculate and predict the output of the machine tools and efficiency of its use. For the theoretical analysis, the mathematical model of productivity rate for a single machine tool with simple technological process is considered.

All manufacturing processes are conducted in space and time. Production time is a basic attribute in mathematical modelling of industrial systems and includes several components, namely time spent on work processes, auxiliary processes, repair and maintenance, and managerial and organisational activity. Engineering industries need holistic mathematical models of time spent in manufacturing processes, where production time should be considered and based on work time and factors that reduce efficiency of time use. Manufacturers need clear mathematical models that enable for computing the actual productivity rate of industrial machines and systems, and the factors that will reduce output. These models of productivity rate are derived according to the level of consideration of the manufacturing processes. Such holistic models for productivity rate should contain all components of production systems activity: the technological and technical indices of the manufacturing systems, the maintenance of machines and repair time of random failures, managerial and organisational activities.

The production time that is spent on manufacturing processes should be evaluated by attributes that can give a clear picture of how time is used. Actual production processes depend on many factors, including technology, machinery design, its exploitation and service systems, the reliability and maintenance of machinery, and managerial and organisational aspects of production processes. Hence, production processes based on the concerted activity of different industrial departments and divisions enable using the production time effectively. However, the real world of industrial environments demonstrates that the production time is not mostly used fully during the manufacturing processes.

The mathematical models present the conceptual principles of productivity theory for production times. Analysis of actual production processes of any machines or systems demonstrates that some typical components of time are permanent in manufacturing processes, and thus these can be mathematically formulated. Analytically, the production time is represented by the following expression: $\theta = \theta_w + \theta_i$, where θ is the total production time; θ_w is the work time and θ_i is the idle time of a machine or system.

Productions processes are evaluated by different indices and one of them is the productivity rate of the machine or system that is represented by the following equation in scale of the total production time:

$$Q = \frac{z}{\theta} = \frac{z}{\theta_w + \theta_i} \tag{2.1}$$

where Q is the actual productivity rate of a production machine or system, z is the number of products manufactured per the total production time θ, and other components are as specified earlier.

The work time θ_w (Eq. 2.1) of a machine or system can be formulated by the following expression: $\theta_w = Tz$, which is a product of a cycle time and number of work parts manufactured per work time θ_w. The cycle time T is represented as a repeatable time for manufacturing of one work part among others, which have the same design and technology of machining. Graphically, the cycle time T for machining of the work parts is presented in Figure 2.5, where the cycle time contains the machining time t_m that spends on machining process (turning, drilling, boring, grinding, assembly, etc.) and the auxiliary time t_a. Later one spends on load and clamp of a work part at a machining area, fast motions of the tools forward to a work part, etc., before machining processes and for fast motions back of the tools from a machined work part, release and remove a work part from machining area, etc., after ending of machining process. Following motions at the time are repeatable for other work parts n. So, the cycle time is represented by the following expression: $T = t_m + t_a$.

The manufacturing processes are implementing at time and space. The components of time spent on a discrete manufacturing process can be considered by following examples of machining work parts. The machining time of the one work part t_m by the single tool is formulated by the equations that depend on the type of processing like turning, milling, drilling, grinding, assembly etc. (Figure 2.6).

For example, the machining time for the turning operation is calculated by the following equation $t_m = \pi D(l + 2s)/Vs$ and for milling operation by

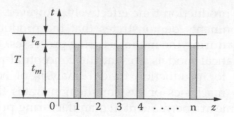

Figure 2.5 Cycle time and its components.

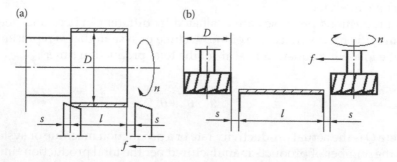

Figure 2.6 Scheme of machining by turning (a) and milling (b) operations.

the equation $t_m = (l + 2s + D)/f$, where t_m is the machining time; f is the feed rate; l is the length of the work part's surface to be machined; s is the safety distance that is necessary to avoid the occasional hits of the cutter to the part at the beginning of machining process and to finish machining the surfaces with guarantee; D is diameter of the shaft for turning operation and diameter of the milling cutter; V is the cutting speed; n is the speed of the rotation for the shaft and the milling cutter. The expressions of the machining time for the other operations can be formulated by a similar way.

For a production system, it is important to use the production time in full scale and to not have delays or idle times in manufacturing processes. From this point of view, the cycle time T contains the machining time t_m that is productive time and the auxiliary time t_a that is non-productive time and represents the time loss. A big value of an auxiliary time increases the value of the cycle time and overall production time. This is the reason that the value of the auxiliary time should be short when compared with the machining time. Manufacturers design productive machines with minimal auxiliary time for discrete machining processes that fulfilled the supplementary mechanisms. Designers of the industrial machines strive to decrease the auxiliary and machining times also with the aim to get their high output. Generally, the value of auxiliary time t_a expresses the dynamics of auxiliary mechanisms and quality of a machine design. However, decrease in an auxiliary time is restricted by dynamical characteristics and reliability of machine work.

At manufacturing industries, there are quite many machine tools designed for continuous machining processes where the auxiliary time is absent, $t_a = 0$. This peculiarity of such machine tools represents a machine design index. Such machine is considered as perfectly designed and no time losses exist on any auxiliary motions, e.g. loading, clamping, reloading, of work parts. The cycle time of machine tools without the auxiliary motions for the continuous manufacturing process is represented by the following expression, $T = t_m$ and the productivity rate is $Q_t = 1/t_m$ that is the maximal for a given technology. Nevertheless, decrease of machining time on machine tools is limited due to the physical process of technology, reliability of a machine work and cutters. Practically, there are several restrictions due to technological and technical factors, and there are optimal parameters of a machine work that enable defining the predictable maximal productivity for the machine.

The idle time θ_i of a production process contains the following main components: $\theta_i = \theta_r + \theta_m$. where the first θ_r is the time spent on repair, maintenance and service of random machinery failures due to many technical reasons. In addition, this θ_r includes the time spent on fabricating the defected products that is a result of non-reliable work of a control system and non-reliable technology. The second component θ_m is the idle time that expresses the quality of management and organisations systems, which spends on two types of downtimes:

• Unpredictable interruptions in a production process due to the absence of work parts, power drops, employee delay, etc.
• Planned maintenance and service time for a stopped machine, according to the service plan that is needed to keep the machine in working condition for the long run, and the organisational changes needed to prepare the manufacturing process for new design products.

Largely, several factors and two aspects affect a machine's idle times in particular. First, a machine can be interrupted due to random reasons, and second, a machine can be stopped for planned reasons according to the machine's preventive maintenance and service processes. The second factor is considered as a management and organisational process, which is evaluated as the production system efficiency. For the following analysis of the productivity rates of machines, it is necessary to separate the repair time for the random stops, and the repair time for the planned preventive stops. However, these two types of times for machine repair are components of a management activity that should be presented in one mathematical model.

Substituting defined parameters into Eq. (2.2) and transformation yield the following equation [14,17]:

$$Q = \frac{z}{\theta_w + \theta_r + \theta_m} = \frac{1}{T + (\theta_r/z) + (\theta_m/z)} \qquad (2.2)$$

where all components are as specified earlier.

The components of Eq. (2.2) should be considered in detail. The component of idle time θ_r is related to the reliability of a manufacturing machine. Expression θ_r/z can be formulated by indices of a machine reliability, which are the machine failure rate λ, mean time of work $m_w = 1/\lambda$ and mean time of repair m_r. Then, the expression $\theta_r = m_r n$, where n is the number of random failures of a machine and $z = \theta_w/T = m_w n/T$, where all components are as specified earlier. Substituting defined components and transformation yield the following equation:

$$\frac{\theta_r}{z} = \frac{m_r n}{m_w n/T} = m_r \lambda T$$

The component of idle time θ_m is related to managerial and organisational problem of a production process and expression θ_m/z is formulated by the following equation:

$$\frac{\theta_m}{z} = \frac{\theta_m}{\theta_w/T} = \frac{\theta_m T}{\theta_w}$$

where all components are as specified earlier.

Substituting defined parameters into Eq. (2.2) and transformation yield the following equation for the actual output of production process:

$$Q = \frac{1}{T + m_r \lambda T + (\theta_m/\theta_w)T} = \frac{1}{T} \times \frac{1}{1 + m_r \lambda + (\theta_m/\theta_w)}$$

$$= \frac{1}{t_m + t_a} \times \frac{1}{1 + m_r \lambda + f_m} \qquad (2.3)$$

where $f_m = \theta_m/\theta_w$ is the management index of production system.

Defined Eq. (2.3) of actual productivity rate for a production system contains the technological, technical, reliability and managerial parameters that enables for evaluation of the all types of productivity rate separately. Equation (2.3) represents the holistic conceptual mathematical model of the productivity rate that includes all components, which have direct links and influence on the production process output.

Production processes are complex systems that depend on many factors as previously mentioned. It is necessary to consider the different types of productivity rates of a production system to allow for an evaluation of the efficiency of its components. Such an approach makes it possible to separately consider the technology of the process, the design and reliability of the machine, and the efficiency of the management and organisation systems of production processes. Analysis of the productivity rate of an industrial machine or system is implemented by considering two components, namely productive and non-productive times that spend on a manufacturing process. Any components of a production time that are not linked with a manufacturing process are considered as production time losses. In such cases, the known expression 'time is money' is a right manifestation of the economic principle of a production process.

Having confirmed the link between the technology t_m, design quality t_a, reliability indices $m_r\lambda$ and the management index f_m (Eq. 2.3), the next step in the analysis of the production system is to focus on the weight of each component on actual productivity rate. The management and organisational index (for simplicity, this shall be referred to as the management index) in the production system play a big role in its productivity. However, the management index f_m is not a direct part of the manufacturing process and does not relate to technological and technical parameters of the manufacturing machine or system, but is rather linked to the availability of a production system. This component is omitted from Eq. (2.3) for the following considerations of the indices of productivity rate for manufacturing machines and systems. However, for general analysis of the efficiency of production systems, the management index f_m can be included for consideration.

For evaluation of the productivity rate of a manufacturing process only, the management index f_m is removed from Eq. (2.3). Then, the productivity rate of a manufacturing machine is expressed by the following equation [14,17]:

$$Q_m = \frac{1}{t_m + t_a} \times \frac{1}{1 + m_r\lambda} \tag{2.4}$$

where $A = 1/(1 + m_r\lambda)$ is the availability of a manufacturing machine, and other components are as specified earlier.

The mathematical model for productivity rate (Eq. 2.4) contains technological, design and reliability parameters of a single machine tool. As mentioned earlier any industrial machine consists of different components and units with different indices of reliability. The machine tool can consist of the following mechanisms and units: spindle, supports, cutters, gearings, electric, electronic, pneumatic or hydraulic units, etc. The modified Eq. (2.4) is expressed as follows [16]:

$$Q = \frac{1}{t_m + t_a} \times \frac{1}{1 + m_r \sum_{i=1}^{n} \lambda_i} \tag{2.5}$$

where $\lambda = \sum_{i=1}^{n} \lambda_i$ is the failure rate of one single machine or workstation, λ_i is failure rate of a mechanism i and other parameters are as specified earlier.

For a machine with ideal reliability, i.e. $A = 1$, the productivity rate is represented by the following equation:

$$Q_c = \frac{1}{t_m + t_a} \tag{2.6}$$

which express the output of a machine per cycle.

For a machine with ideal design and ideal reliability, i.e. $t_a = 0$ and $A = 1$, the productivity rate is represented by the following equation:

$$Q_t = \frac{1}{t_m} \tag{2.7}$$

which expresses the maximal output of a machine tool with continuous technological process.

In manufacturing industries, there are many designs of machine tools with continuous technological process. Typical examples can be represented by the following constructions:

- Wire stripping, crimpling automatic machine.
- Thread rolling automatic machine.
- Nut tapping threading automatic machine.
- Cylindrical grinding automatic machine, etc.

These automatic manufacturing machines are designed with continuous technological processes without auxiliary motions for loading and reloading work parts. This construction of manufacturing machines is considered as a perfect design for the given technological process.

A working example

Productivity rates of manufacturing system and the factors that lead to decreasing production system outputs are shown in Eqs. (2.1–2.6). Given later is an analysis of the productivity rates and factors for a typical production system.

Table 2.1 Technical data of a production system

Title	Data
Machining time, t_{mo} (min)	5.0
Auxiliary time, t_a (min)	0.5
Failure rate of machine units, λ (per min)	1.0×10^{-2}
Mean repair time, m_r (min)	3.0
Production time, θ_p (min)	420.0
Idle time, θ_i (min)	20.0

Table 2.2 The productivity data for the production system

Productivity rate		Productivity losses due to	
Q_t (product/h)	12	Auxiliary time	$\Delta Q_1 = 1.14$ (product/h)
Q_c (product/h)	10.86	Availability, $A = 0.970$	$\Delta Q_2 = 0.36$ (product/h)
Q_m (product/h)	10.5	Management index	$\Delta Q_3 = 42$ (product/h)
Q_a (product/h)	10.08	Total losses	$\Delta Q = 1.92$ (product/h)

An engineering problem is presented, and the productivity rates and factors are calculated for the given data of the manufacturing system, shown in Table 2.1, where the technical and timing data of the production system refer to one product.

These results are obtained by substituting the initial data (Table 2.1) into Eqs. (2.1–2.6). The calculated results represented in Table 2.2 that enable to facilitate the evaluation of the attributes of the productivity rates for the manufacturing system.

The technological productivity rate is as follows:

$$Q_t = \frac{1}{t_m} = \frac{1}{5.0} = 0.2 \quad \text{product/min} = 12.0 \quad \text{product/h.}$$

The productivity rate per cycle time is as follows:

$$Q_c = \frac{1}{t_m + t_a} \times \frac{1}{5.0 + 0.5} = 0.181 \ \text{product/min} = 10.86 \ \text{product/h.}$$

Productivity losses due to the auxiliary time is

$$\Delta Q_1 = Q_t - Q_c = 0.2 - 0.181 = 0.019 \ \text{product/min} = 1.14 \ \text{product/h.}$$

The machine productivity is as follows:

$$Q_m = \frac{1}{t_m + t_a} \times \frac{1}{1 + m_r \lambda} = 0.181 \times \frac{1}{1 + 3.0 \times 1.0 \times 10^{-2}}$$

$$= 0.175 \ \text{product/min} = 10.5 \ \text{product/h.}$$

where the availability is $A = \dfrac{1}{1+3.0\times1.0\times10^{-2}} = 0.970$

Productivity losses due to the machine availability is as follows:

$\Delta Q_2 = Q_c - Q_m = 0.181 - 0.175 = 0.006 \text{ product/min} = 0.36 \text{ product/h}.$

The production system actual output is as follows:

$$Q_a = \frac{1}{T}\times\frac{1}{1+m_r\lambda+f_m} = \frac{1}{5.5}\times\frac{1}{1+3.0\times1.0\times10^{-2}+20/400}$$

$$= 0.168 \text{ product/min} = 10.08 \text{ product/h}.$$

where the management index is $f_m = \theta_m/\theta_w = \theta_m/(\theta - \theta_i) = 20/(420 - 20) = 0.05$ (Table 2.1).

Productivity losses due to the management index is as follows:

$\Delta Q_3 = Q_m - Q = 0.175 - 0.168 = 0.007 \text{ prod/min} = 0.42 \text{ product/h}.$

Graphical presentation for the productivity rates of a manufacturing system is shown in Pareto chart (Figure 2.7, Table 2.2). Pareto chart helps to analyse the data easily and provide a clearer view of results and priority directions for sustainable improvement of the production system. The productivity rate diagram enables illustrating and summarising the results in the bar line chart, Figure 2.7. This chart shows a decrease of productivity rate from potential productivity outputs with regard to the actual outputs due to design, reliability and management reasons.

The values of productivity losses by other reasons are relatively less. Figure 2.7 is actually the summary of the calculation performed earlier. If the production manufacturing system is 100% efficient, it can produce 12 products per hour. However, due to several losses of productivity rates during the process, it actually produces only $Q_a = 10.08$ products per hour. The production system has 97.0% of its efficiency. Conducted calculations demonstrate that attributes of productivity rate contain typical and real components of technical and technological parameters for manufacturing processes. These attributes enable evaluating the process and the machine tool from point of appropriateness and effectiveness of the company.

Figure 2.7 clearly demonstrates the maximal, possible and actual productivity rates and reasons of productivity losses for the given production system. This information is represented using the visual example for the following analysis of production system with an aim to decrease its losses and enhance its efficiency. All productivity losses should be considered as the sources of productivity improvement for the production system by decreasing defined losses. Each type of productivity loss should be considered in detail by technology and design departments. For example, the productivity loss due to reliability

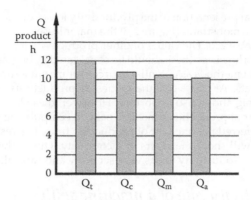

Figure 2.7 Productivity diagram parameters Q_t, Q_c, Q_m and Q_a of the manufacturing system.

problem contains different units in design, i.e. mechanical, electric, electronic, etc. Hence, they should be considered separately. Detailed analyses of reasons of productivity losses and the following engineering solutions enable to increase the productivity rate and efficiency of the production system. Figure 2.7 is the typical diagram of an actual output of production systems, but productivity losses for different industries are different. For example, in metal-cutting processes most of the productivity losses fall to auxiliary time and reliability of tools and units. In an electronic industry, productivity losses fall to defected products due to reliability and quality of technological processes.

The values of productivity rate losses are observed clearly through the Pareto chart (Figure 2.8). It helps to analyse the results and comes with priority directions of sustainable improvement of the productivity rate of manufacturing system.

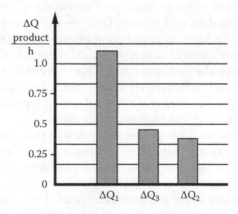

Figure 2.8 Losses of productivity rate of a manufacturing system.

Graphical presentation of the productivity losses of a manufacturing system demonstrates (Figure 2.8) the major losses due to auxiliary time ΔQ_1 (Table 2.2). The values of other productivity losses (ΔQ_3 and ΔQ_2) are less; however, these should be considered by engineering and management methods as potential sources for manufacturing system improvements. At first sight, the values of productivity losses look like not so big and not so important. However, these losses referred per one hour of production time. It is easy to compute the productivity losses referred per one year of production time. The result of such computing will shock managers of a company. This is the reason to consider the productivity losses very scrutiny and carefully.

2.4 Productivity rate of a machine with multi-tool machining process

Manufacturers are struggling to increase the productivity rate of machining processes and to put more products on the market. Increase in the productivity rate is achieved by intensification of the machining processes, which leads to an increase in the costs of machining processes and that of products. The two most important attributes for the economics of manufacturing are the minimum machining cost per part and the maximum production rate of machine tools. Which of these attributes are more important depends on several factors. The marketing process is regulating this process. If customers request a product, manufacturers can sell at high prices. In this case, the attribute of minimum machining cost moves away from its first position while the attribute of the high productivity rate by intensification of machining process comes to the first level. When a market comes to be saturated with products, the sales of a product decrease, in such a case, the product should be manufactured using the minimum machining cost, which method is well described in the textbooks of economics.

In the past, both manufactures and researchers have collected significant amounts of data and information regarding the reliability attributes of different machine units and components. This valuable data have allowed for the development of mathematical models for the productivity rate of different manufacturing machines and systems, in terms of technical, technological and reliability attributes. Such an approach produces a simple analytical presentation of the manufacturing processes that is both understandable and convenient in calculations.

The desire to increase the productivity rate of machine tools has led to strategies that involve intensifying the manufacturing processes and finding mathematical models for the maximal productivity rate of machine tools. The optimisation of a machining process using the criterion of maximum productivity rate is therefore crucial. This problem is complex and, in many cases, contains unresolved questions connected with multivariable parameters of manufacturing processes. Modern manufacturing

(a)   (b)

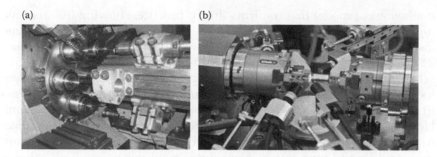

Figure 2.9 Multi-tool machining process at the multi-spindle machine tool (a) and at the single spindle machine tool (b).

processes use different technologies with application of single and multi-tool machining processes. Multi-tool processes are conducted in multi-spindle and multi-station machine tools, turret-type lathes, automated lines, etc. (Figure 2.9). Tools are engaged simultaneously, in such circumstances, and to define the optimal machining parameters for different tools by criterion of the maximal productivity rate is not simple problem. Intensification of machining processes by the simultaneous action of different tools leads to non-proportional change in the productivity rate.

The mathematical model of productivity rate that contains the reliability indices of machine tool units is the subject of special investigation. On the one hand, intensification of machining processes in industrial machine tools leads to an increase in their productivity rate; on the other, it results in an increase in the failure rates of the same machine tool units. The increase in dynamic loads on machine units lead to conditions (e.g. increases in the wearing process, vibrations in the machine) that result in machine tool downtimes, and hence decrease in productivity rates. In such circumstances, a mathematical model for the productivity rate that contains changing machining parameters allows for the discovery of optimal cutting regimes giving a maximal productivity rate.

In the manufacturing area, machine tool's units and mechanisms have different indices of reliability that mainly manifested the failure rates. Weakest element of machine tools is cutters relative to failure rates of other mechanisms and units. The cutters are replaced in machine tools in average after 1 or 2 hours of work, which depends on the type of cutter material. Other primary machine units like spindles, supports, gearboxes, mechanisms of machining motions, etc. have reliability indices, which values many times higher than the reliability indices of the cutters. Practice demonstrates that intensification of machining regimes do not reflect too much on the change in the reliability indices of machine tool mechanisms. This result is based on acquired data that the failure rate of cutters is prevalent many times over the failure rates of other machine units. There is a mathematical model that confirms this in practice, showing that increasing the normative cutting speed 4–5 times leads to an increase in the time losses due to the reliability of the main machine units at no more than 2%. In such circumstances, time losses due to machine units do not give sensitive results relating to drop in the productivity rate of the machine tool. Hence, the time losses of primary machine tool units with changes in the machining regimes for common cases can be accepted as constant. However, it is necessary to point that the modern tendency in manufacturing areas is to use new types of cutter materials, whose properties are close to diamond-type cutters. New cutter materials have a high value of reliability that can be commensurable to the reliability value of primary machine units. In such cases, deriving the mathematical model of productivity rate for the machining process implemented on a machine tool that includes failure rates of other units and mechanisms would be crucial.

The criterion of maximum productivity rate for the machinery prevails in many cases of industrial production. Practice demonstrates no great difference in the values of optimal machining parameters calculated by the criteria of minimum cost and maximum productivity rate. Analytically, unsolved problems of machining parameters' optimisation are corrected practically through the path's long process of analysis by both criteria. The lack of mathematical models among numerous publications, capable of giving correct results for the optimisation of machining parameters, shows that this difficult problem is still a debated topic.

Mathematical models of optimal cutting speed for multi-tooling machining processes of simultaneous and separate actions are presented later. These models describe changes in machining time and time losses due to cutter's replacement and tuning and time losses of main machine tool mechanisms and units. Mathematical model derived is universal and can be applied to multi-spindle, multi-station machine tools and automated lines with multi-tooling machining processes.

Mathematical models of machining regime optimisation are considered by the maximal productivity rate criterion [15]. The cutting speed, feed rate and depth of the cut are parameters of a machining regime. Feed rate and depth of the cut cannot be changed in large-scale operations due to limitations in the quality of the surface and the accuracy of the machining process. Hence, the feed rate and depth of cut is accepted as constant in mathematical models, because they do not influence the failure rate of cutters with intensification of machining processes. The mathematical model for productivity rate does not consider the change of the tool life from parameters of feed rate and depth of cut. However, the cutting speed can be changed over quite a sizable range and its increase often leads to improvements in both the quality and productivity in the machining process. On the other hand, increasing the cutting speed reduces the cutter's tool life and more intensively leads to its replacement, and increasing the downtime due to cutters reduces the productivity rate. The action of all pointed factors is represented in a mathematical model for productivity rate as a function of intensification of machining process.

Based on studies of productivity presented earlier, the productivity rate of a machine tool in terms of machining and auxiliary times, as well as reliability attributes, is presented by the following equation:

$$Q = \frac{1}{t_{mn} + t_a} \times \frac{1}{1 + m_r \left(\sum_{i=1}^{n} \lambda_{c.i} + \sum_{i=1}^{k} \lambda_{s.i} + \lambda_{cs} + \lambda_g \right)} \tag{2.8}$$

where t_{mn} is normative machining time; λ_i is failure rate of i mechanisms and units (cutter λ_c, spindle λ_s, control system λ_{cs}, gearing λ_g, etc.) of the machine tool; other parameters are as specified earlier.

The industrial practice and theory of machining processes demonstrate that an increase in cutting speed leads to a decrease in machine time, as well as an increase in the failure rate of machine tool components. It is important to consider each component of the mathematical model for a productivity rate in terms of the change in cutting speed of the machine tool. Hence, new machining time for processing a work part changes in proportion to changes in the cutting speed and is represented by the following equation:

$$t_m = \frac{t_{m.n}}{c} = \frac{t_{m.n}}{V/V_n} = \frac{t_{m.n} V_n}{V} \tag{2.9}$$

where t_m is the new machining time for the longest operation, $c = V/V_n$ is the coefficient of change in the cutting speed, V is the new cutting speed and V_n is the normative cutting speed.

The change in auxiliary time relative to changes in the machining process depends on the design of the machine tool. Most modern designs of the machine tools do not have kinematic links between mechanisms of machining and auxiliary motions. Auxiliary motions, such as the work-part feeding, fast motions of the tool to the machining area and back, handling motions, etc., are not part of the machining process. Hence, for most types of machine tools, the auxiliary time is outside the machining process and changes in the machining process do not reflect on the auxiliary motions' time. Based on these circumstances, the time of auxiliary motions is accepted as constant for manufacturing machines, i.e. $t_a = \text{const}$.

In the case where machine tool designs have hard kinematical links between mechanisms of machining and auxiliary motions, the auxiliary time has the proper function of change in the machining process. Most of these types of machine tools are rare designs, and their number used in the manufacturing area is small and does not play a big role. So, for further analysis, these types of machine tools are not considered.

The change in failure rates due to the reliability of cutters has a mathematical function. The change in the machining process leads to changes in the failure rates of cutters. The mathematical equation for the increase in failure rate due to an intensive wear process and reliability of cutting tool with an increase in cutting speed can be found via the following approach.

The equation for the increase in the failure rate of cutters with an associated increase in cutting speed is found using the equation of the Taylor tool life, $T_n = \left(C/V_n\right)^{1/b}$, where C is the empirical constant resulting from regression analysis and field studies, which depends on many factors (geometry of the tool, cooling process, cutting speed, feed rate, depth of cut, surface hardness of the workpiece, etc.). V_n (m/min) is the normative cutting speed of a machining process and b is the empirical constant that generally depends on the cutter tool material.

The attribute λ_c of cutter failure rate is changed along with any change in the machining process. This change is a subject for mathematical analysis. This attribute, the cutter failure rate at normative machining process, has the following equation:

$$\lambda_{c.n} = \frac{1}{m_w} = \frac{1}{T_n} \tag{2.10}$$

where m_w is the mean time of cutter work between two replacements and T_n is the normative tool life ($m_w = T_n$ is evident).

Substituting the Taylor tool-life equation into Eq. (2.10) yields the following equation:

$$\lambda_{c.n} = \frac{1}{\left(C/V_n\right)^{1/b}} = \left(\frac{V_n}{C}\right)^{1/b} \tag{2.11}$$

$V = V_n c$ expresses the change in the cutting speed and the tool-life equation at a new condition is represented by the following equation:

$$T_c = \left(\frac{C}{V_n c}\right)^{1/b} \tag{2.12}$$

Substituting defined parameters into Eq. (2.11) and transforming give the equation of the failure rate of the one cutter at the new condition as follows:

$$\lambda_c = \frac{1}{\left(\dfrac{C}{V_n c}\right)^{1/b}} = \frac{\left(\dfrac{V}{V_n}\right)^{1/b}}{\left(\dfrac{C}{V_n}\right)^{1/b}} = \left(\frac{V}{C}\right)^{1/b} \tag{2.13}$$

The failure rates of n independent cutters involved in machining process according to the rules of probability theory are represented by the following equation:

$$\sum_{i=1}^{n} \lambda_{c.i} = \sum_{i=1}^{n} \left(\frac{V_i}{C_i} \right)^{1/b_i} \tag{2.14}$$

Equation (2.14) should be considered for the two types of machining processes. In case of the single spindle machine tool or multi-spindle machine tool with one speed of spindle's rotation, the cutting speed is equal to all n cutters involved in machining process simultaneously. For the machine tools with independent cutting speeds of the tools, Eq. (2.14) contains different n cutting speeds that should be represented proportional to the cutting speed of the longest operation.

This method enables to have only one basic variable parameter in the following analysis. In such case, the failure rates of the cutters are represented as follows:

$$\sum_{i=1}^{n} \lambda_{c.i} = \sum_{i=1}^{n} \left(\frac{a_i V}{C_i} \right)^{1/b_i} \tag{2.15}$$

where $a_i = V_i/V$ is the ratio of the cutting speed of a cutter i relative to the cutting speed of the cutter with longest machining time.

Substituting defined parameters and Eqs. (2.9) and (2.15) into Eq. (2.8) and the following transformation yield the following equation for the productivity rate of the multi-tool machining process with variable cutting speed:

$$Q = \frac{1}{\frac{t_{m.n} V_n}{V} + t_a} \times \frac{1}{1 + m_r \left[\sum_{i=1}^{n} \left(\frac{a_i V}{C_i} \right)^{1/b_i} + \sum_{j=1}^{r} \lambda_j \right]} \tag{2.16}$$

where all parameters are as specified earlier.

Analysis of Eq. (2.16) shows that two variables are changed with increase of cutting speed V. The first is machining time $t_{m.n} V_n/V$ that is decreasing, and the second is the failure rate of cutters $\sum_{i=1}^{n} (a_i V/C_i)^{1/b_i}$, whose magnitude is increasing. Hence, Eq. (2.16) has an extreme function, i.e. there is optimal change in the cutting speed V_{opt} that gives maximal productivity rate to a machine tool.

The extreme function of Eq. (2.16) can be defined by the rules applied for differential equations. Figure 2.10 depicts the change in variable components of the Eq. (2.16) versus the change in the cutting speed V.

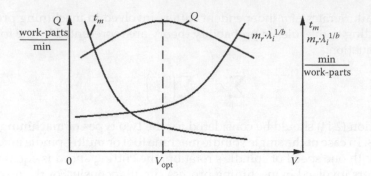

Figure 2.10 Change of the machine productivity rate parameters versus change of the cutting speed.

The optimal cutting speed using the maximal productivity rate criterion for longest operation is obtained by the first derivative of Eq. (2.16) with respect to the cutting speed and setting it to zero. Then, the first derivative gives the following differential equation:

$$\frac{dQ}{dV} = \frac{d\left\{\dfrac{1}{\dfrac{t_{m.n}V_n}{V}+t_a}\times\dfrac{1}{1+m_r\left[\left(\dfrac{a_1V}{C_1}\right)^{1/b_1}+\left(\dfrac{a_2V}{C_2}\right)^{1/b_2}+\cdots+\left(\dfrac{a_nV}{C_n}\right)^{1/b_n}+\sum\limits_{j=1}^{r}\lambda_j\right]}\right\}}{dV}=0$$

giving rise to the following

$$\frac{t_{m.n}V_n}{V^2}\left\{1+m_r\left[\left(\frac{a_1V}{C_1}\right)^{1/b_1}+\left(\frac{a_2V}{C_2}\right)^{1/b_2}+\cdots+\left(\frac{a_nV}{C_n}\right)^{1/b_n}+\sum_{j=1}^{r}\lambda_j\right]\right\}$$

$$-\left(\frac{t_mV_o}{V}+t_a\right)m_r\left[\frac{a_1}{b_1C_1}\left(\frac{a_1V}{C_1}\right)^{\frac{1}{b_1}-1}+\frac{a_2}{b_2C_2}\left(\frac{a_2V}{C_2}\right)^{\frac{1}{b_2}-1}+\cdots+\frac{a_n}{b_nC_n}\left(\frac{a_nV}{C_n}\right)^{\frac{1}{b_n}-1}\right]=0$$

$$(2.17)$$

Simplification of Eq. (2.17) and transformation yield the following equation:

$$\frac{\dfrac{1}{m_r}+\sum\limits_{j=1}^{r}\lambda_j}{\left(1+\dfrac{t_aV}{t_{m.n}V_n}\right)}=\sum_{i=1}^{n}\left(\frac{1}{b_i}-1\right)\left(\frac{a_iV}{C_i}\right)^{1/b_i} \qquad (2.18)$$

where all parameters are as specified earlier.

Equation (2.18) is transcendental, where roots cannot be found analytically. Solution of a transcendental equation is resolved by using graphical methods. The left and right sides of Eq. (2.18) are exponential functions and can be represented by the following equations:

$$f(V_{opt}) = \frac{\dfrac{1}{m_r} + \sum\limits_{j=1}^{r} \lambda_j}{\left(1 + \dfrac{t_a V}{t_{m.n} V_n}\right)}, \quad g(V_{opt}) = \sum_{i=1}^{n}\left(\frac{1}{b_i} - 1\right)\left(\frac{a_i V}{C_i}\right)^{\frac{1}{b_i}} \tag{2.19}$$

The change of argument V of these functions with given data lead to a decrease of the function $f(V_{opt})$ and to increase of the function $g(V_{opt})$. The intersection of functions $f(V_{opt})$ and $g(V_{opt})$ gives the solution of the optimal cutting speed. Equation (2.19) can be solved manually.

A working example

The multi-tool simultaneous machining process is represented in Figure 2.11. The carbon steel work part is machined by the boring cutter, drill bit and milling tool simultaneously on the machine tool. The machining parameters and properties of the machine tool and cutters are represented in Table 2.3 [1].

The cutting speed for all tools should be optimised using the criterion of maximum productivity rate.

Solution: The analysis of machining parameters and operations represented in Table 2.3 show that the cutting speeds for machining different surfaces of the work part are different. The longest machining time is conducted by the milling tool $t_m = 5.0$ min, which is accepted as basic. The optimal cutting speed using the criterion of maximum

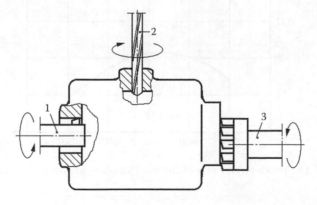

Figure 2.11 The work-part's surfaces machined by the three tools simultaneously.

Table 2.3 Machining and tool's normative parameters

Tool	Cutter 1, boring	Drill bit 2	Milling tool 3
Tool material	Alumina ceramic	High speed steel	Cemented carbide
Machining time, $t_{m.i}$ (min/part)	3.0	4.0	5.0
Auxiliary time, t_a (min/part)	0.3		
Cutter's constant $\quad C_i$	400	30	150
$\qquad\qquad\qquad b_i$	0.5	0.125	0.25
Tool life, T_o (min)	100	45	70
Normative cutting speed, V_{io} (m/min)	200	24	80
Ratio of the cutting speeds, a_i	200/80 = 2.5	24/80 = 0.3	1
Failure rate of machine units, $\sum\limits_{j=1}^{r}\lambda_j$	6×10^{-7}		
Mean repair time, m_r (min)	2.0		

productivity rate for multi-tool machining processes is defined by Eq. (2.19). Substituting the given parameters (Table 2.3) into Eq. (2.19) and calculating represent the result in Figure 2.12, which is the diagram of change in the productivity rate of multi-tool machining processes versus the change in cutting speeds.

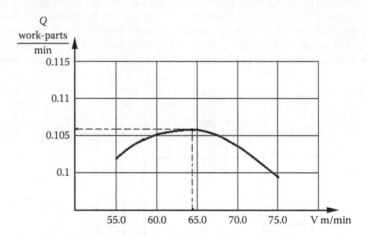

Figure 2.12 The productivity rate of the multi-tooling process versus the cutting speed.

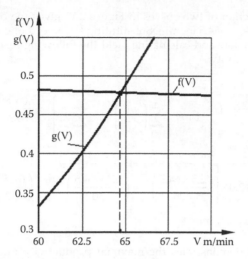

Figure 2.13 Graphical solution of the optimal cutting speed for the multi-cutting process.

Equation (2.19) defines the optimal cutting speed of the basic milling tool and gives the maximum productivity rate, and is represented graphically in Figure 2.13.

$$f(V_{opt}) = \frac{\dfrac{1}{m_r} + \sum_{j=1}^{r} \lambda_j}{\left(1 + \dfrac{t_a V}{t_m V_o}\right)} = \frac{\dfrac{1}{2.0} + 6.0 \times 10^{-7}}{\left(1 + \dfrac{0.3 \times V}{5.0 \times 80.0}\right)}$$

$$g(V_{opt}) = \sum_{i=1}^{n} \left(\frac{1}{b_i} - 1\right)\left(\frac{a_i V}{C_i}\right)^{\frac{1}{b_i}} = \left(\frac{1}{b_1} - 1\right)\left(\frac{a_i V}{C_1}\right)^{\frac{1}{b_1}} + \left(\frac{1}{b_2} - 1\right)\left(\frac{a_2 V}{C_2}\right)^{\frac{1}{b_2}}$$

$$+ \left(\frac{1}{b_3} - 1\right)\left(\frac{a_3 V}{C_3}\right)^{\frac{1}{b_3}}$$

$$= \left(\frac{1}{0.5} - 1\right)\left(\frac{2.5 \times V}{400}\right)^{\frac{1}{0.5}} + \left(\frac{1}{0.125} - 1\right)\left(\frac{0.3 \times V}{30}\right)^{\frac{1}{0.125}}$$

$$+ \left(\frac{1}{0.25} - 1\right)\left(\frac{V}{150}\right)^{\frac{1}{0.25}}$$

The intersection of two curves in Figure 2.13 gives the optimal cutting speed $V_{opt} = 64.5$ m/min. Substituting this speed into equation of productivity rate and calculating yield the following result of maximum productivity:

$$Q_{max} = \cfrac{1}{\cfrac{t_{m.3}V_{3.o}}{V}+t_a} \times \cfrac{1}{1+m_r\left[\sum_{i=1}^{n}\left(\cfrac{a_iV}{C_i}\right)^{1/b_i}+\sum_{j=1}^{r}\lambda_j\right]} = \cfrac{1}{\cfrac{5.0\times80}{64.5}+0.3}$$

$$\times \cfrac{1}{1+2.0\times\left[\left(\cfrac{2.5\times64.5}{400}\right)^{\frac{1}{0.5}}+\left(\cfrac{0.3\times64.5}{30}\right)^{\frac{1}{0.125}}+\left(\cfrac{64.5}{150}\right)^{\frac{1}{0.25}}+6.0\times10^{-7}\right]}$$

$$= 0.106\,\text{part/min}$$

Figure 2.12 demonstrates the maximal productivity rate $Q = 0.106$ part/min, which is obtained by the optimal cutting speed for the milling tool $V = 64.5$ m/min. This data enables to calculate the optimal cutting speeds for other cutters using the equation $V_i = a_iV$ and the new machining times for each cutter by Eq. (2.9).

• For the boring cutter 1: $V_1 = a_1V = 2.5 \times 64.5 = 161.25$ m/min;

$$t_{m.1} = \frac{t_{m.o}V_0}{V} = \frac{3.0\times200.0}{161.25} = 3.72\,\text{m/min}$$

• For the drill bit 2: $V_2 = a_2V = 0.3 \times 64.5 = 19.35$ m/min;

$$t_{m.2} = \frac{t_{m.o}V_0}{V} = \frac{4.0\times24.0}{19.35} = 4.96\,\text{m/min}$$

• For the milling tool 3: $V_3 = a_sV = 1.0 \times 64.5 = 64.5$ m/min;

$$t_{m.1} = \frac{t_{m.o}V_0}{V} = \frac{5.0\times80.0}{64.5} = 6.20\,\text{m/min}$$

This example shows that decrease in the cutting speed, which is optimal, leads to an increase in the productivity rate, which is maximal.

Formulated earlier, the mathematical model of productivity rate enables to define the optimal cutting speeds for separate operations that may yield a different productivity rate, which should be compared with the result obtained for multi-cutting process of simultaneous action.

The optimal cutting speed for single-tool machining processes is defined by Eq. (2.19). Substituting the parameters represented in Table 2.3 into Eq. (2.19) and calculating give the following results:

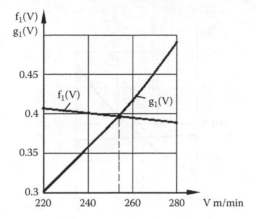

Figure 2.14 Graphical solution of the optimal cutting speed for the boring cutter.

- For the boring cutter 1

$$f_1(V_{opt}) = \frac{\frac{1}{m_r} + \sum_{i=1}^{r} \lambda_j}{\left(1 + \frac{t_a V}{t_m V_o}\right) \times \frac{1}{b_1} - 1} = \frac{\frac{1}{2.0} + 6.0 \times 10^{-7}}{\left(1 + \frac{0.3 \times V}{3.0 \times 200.0}\right) \times \frac{1}{0.5} - 1}$$

$$g_1(V_{opt}) = \left(\frac{V_1}{C_1}\right)^{1/b_1} = \left(\frac{V_1}{400}\right)^{1/0.5}$$

This two equations represent graphically the change of functions $f_1(V)$ and $g_1(V)$ for which the intersection is represented as the optimal cutting speed (Figure 2.14)

The diagram represents the boring cutter in which the optimal cutting speed is 253 m/min.

- For the drill bit 2

$$f_2(V_{opt}) = \frac{\frac{1}{m_r} + \sum_{i=1}^{r} \lambda_j}{\left(1 + \frac{t_a V}{t_m V_o}\right) \times \frac{1}{b_1} - 1} = \frac{\frac{1}{2.0} + 6.0 \times 10^{-7}}{\left(1 + \frac{0.3 \times V}{4.0 \times 24.0}\right) \times \frac{1}{0.125} - 1}$$

$$g_2(V_{opt}) = \left(\frac{V_2}{C_2}\right)^{1/b_2} = \left(\frac{V_2}{30}\right)^{1/0.125}$$

Intersection of functions $f_2(V)$ and $g_2(V)$ is represented in Figure 2.15.

The diagram represents the drill bit in which the optimal cutting speed is 21.3 m/min

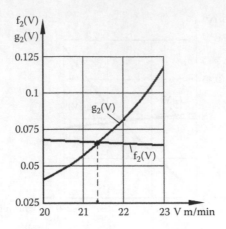

Figure 2.15 Graphical solution of the optimal cutting speed for the drill bit.

- For the milling tool 3.

$$f_3(V_{opt}) = \frac{\dfrac{1}{m_r} + \displaystyle\sum_{i=1}^{r} \lambda_j}{\left(1 + \dfrac{t_a V_3}{t_m V_o}\right) \times \dfrac{1}{b_3} - 1} = \frac{\dfrac{1}{2.0} + 6.0 \times 10^{-7}}{\left(1 + \dfrac{0.3 \times V_3}{5.0 \times 80.0}\right) \times \dfrac{1}{0.25} - 1}$$

$$g_3(V_{opt}) = \left(\frac{V_3}{C_3}\right)^{1/b_3} = \left(\frac{V_3}{150.0}\right)^{1/0.25}$$

Intersection of functions $f_3(V)$ and $g_3(V)$ is represented in Figure 2.16.

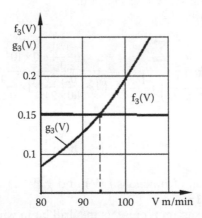

Figure 2.16 Graphical solution of the optimal cutting speed for the milling tool.

The diagram represents the milling tool in which the optimal cutting speed is 94.0 m/min. The optimal cutting speeds for separate cutters can give the new longest operation, which is defined by the proportional change in cutting speed Eq. (2.19). The new cutting speed for boring process gives machining time t_{m1} = 2.371 min/part, for drilling operation t_{m2} = 3.55 min/part and for milling process t_{m3} = 4.255 min/part, which is left as the longest operation. Substituting the defined value of the longest machining time and new optimal cutting speeds for single cutters into Eq. (2.19) and calculating, the result for the maximum productivity rate of machining the workpiece by tools with simultaneous action is as follows:

$$
Q = \frac{1}{\dfrac{t_m V_o}{V} + t_a} \times \frac{1}{1 + m_r \left[\displaystyle\sum_{i=1}^{n} \left(\frac{a_i V}{C_i} \right)^{1/b_i} + \sum_{j=1}^{r} \lambda_j \right]} = \frac{1}{4.255 + 0.3}
$$

$$
\times \frac{1}{1 + 2.0 \times \left[\left(\dfrac{253.0}{400} \right)^{1/05} + \left(\dfrac{21.3}{30} \right)^{1/0.125} + \left(\dfrac{94.0}{150} \right)^{1/0.25} + 6.0 \times 10^{-7} \right]}
$$

$$
= 0.098 \text{ work part/ min}
$$

where $V_{i.opt} = a_i V$, other parameters are as specified earlier.

The productivity rate for the normative machining process by tools with simultaneous action is calculated substituting the parameters represented in Table 2.3 into Eq. (2.19) that give the following result:

$$
Q = \frac{1}{\dfrac{t_m V_o}{V} + t_a} \times \frac{1}{1 + m_r \left[\displaystyle\sum_{i=1}^{n} \left(\frac{a_i V}{C_i} \right)^{1/b_i} + \sum_{j=1}^{r} \lambda_j \right]} = \frac{1}{5.0 + 0.5}
$$

$$
\times \frac{1}{1 + 2.0 \times \left[\left(\dfrac{200.0}{400} \right)^{1/05} + \left(\dfrac{24.0}{30} \right)^{1/0.125} + \left(\dfrac{80.0}{150} \right)^{1/0.25} + 3 \times 2 \times 10^{-7} \right]}
$$

$$
= 0.091 \text{ work part/ min}
$$

The results of productivity rate for the normative and optimal processes for separate and simultaneous actions are presented in Table 2.4.

The analysis of the results demonstrates that the productivity rate of optimal machining process with cutters of simultaneous action (Q = 0.106 work part/min) is higher than for the normative machining

Table 2.4 The productivity rate for normative and optimal cutting processes for single operations and simultaneous action

Cutting speed	Normative for cutters	Optimal for single cutters	Optimal for cutters of simultaneous action
Productivity rate, Q (work part/min)	0.091	0.098	0.106

process ($Q = 0.091$ part/min) and for optimal cutting speeds for single cutters and simultaneous action ($Q = 0.098$ work part/min). The optimisation of machining processes for single operations does not give the maximum productivity rate in the case of simultaneous actions of tools. Equation (2.16) allows for simultaneously minimising the machining time and the failure rates of cutters and optimises their cutting speeds that give maximum productivity rate.

2.4.1 Practical test

Formulated mathematical model of optimisation of the cutting speeds for multi-tool machining process of simultaneous action was practically validated by short tests [15]. Experimental investigation of multi-cutting process conducted on the lathe machine model CY-L1640G with two cutters installed on one tool holder of the support. Two cutters machined the shaft which clamped by the chuck and tailstock. The scheme of turning process by two cutters is represented in Figure 2.17, where the cutters manually set up on the depth of cut t and the high speed steel (HSS) cutter 1 got radial feed and slit the small grove on the rotating shaft. The cemented carbide cutter 2 did not touch the shaft and installed with small safety distance c. The turning process is conducted on the distance$l$by the axial feed rate f of the two cutters.

The actual picture of axial turning process by two cutters is simultaneously represented in Figure 2.18. The data of the cutters, number of their replacements and test machining regimes are represented in Table 2.5.

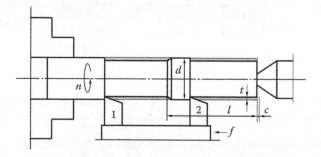

Figure 2.17 Sketch of turning process by two cutters simultaneously.

Figure 2.18 Turning process with two cutters.

Table 2.5 The machining parameters in practical tests

Material of cutter		HSS M42	Cemented carbide insert CNMG-120408
Cutter's index	C	75	400
	b	0.15	0.3
Tool life, T (min)		40	100
Number of cutter replacements, n_r		25	1
The cutter edge wearing (mm)		1.0	1.0
Normative cutting speed, V_o (m/min)		40.0	100
Workpiece material, Shaft diameter, d (mm) AISI 1017 Hardness, HB - 105		25.0	
Cutting speed, V (m/min)		78.539	
Depth of cut, t (mm)		2	
Feed rate, f (mm/rev)		0.153	
Length of machining, l (mm)		110	
Safety distance, c (mm)		2	
Speed of spindle rotation, n (rpm)		1000	
Machining time, t_m (min/work part)		0.738	
Average auxiliary time, t_a (min/work part)		1.27	
Cutter's replacement time, m_r (min/work part)		1.33	
Number of machined work parts, z		200	

Machining process is roughly turning. At the test time, the machine units did not have failures. The data for the tool life of the cutters were obtained from the practical tests and published handbooks, textbooks and research papers. The theoretical productivity rate of machining process with two cutters of simultaneous action is calculated with initial cutting speed $V_0 = 40$ m/min, which correspond to the following number of spindle's revolution:

$$n = \frac{1000V_0}{\pi d} = \frac{1000 \times 40.0}{\pi \times 25.0} = 509.295 \text{ rpm}$$

The machining time is as follows:

$$t_m = \frac{l+c}{fn} = \frac{110+2}{0.153 \times 509.295} = 1.437 \text{ min}$$

where all parameters are as specified in Table 2.5 and Figure 2.17.

Substituting the parameters represented in Table 2.5 into Eq. (2.16) and transforming give the following result:

$$Q = \frac{1}{\dfrac{t_m V_0}{V} + t_a} \times \frac{1}{1 + m_r \left[\left(\dfrac{V}{C_1} \right)^{1/b_1} + \left(\dfrac{V}{C_2} \right)^{1/b_2} \right]}$$

$$= \frac{1}{\dfrac{1.437 \times 40.0}{V} + 1.27} \times \frac{1}{1 + 1.33 \times \left[\left(\dfrac{V}{75} \right)^{1/0.15} + \left(\dfrac{V}{400} \right)^{1/0.3} \right]}$$

The diagram of the change of the productivity of the machine tool versus the cutting speed is represented in Figure 2.19.

Optimal cutting speed for the HSS cutter is calculated by Eq. (2.19). Substituting initial data and transforming yield the following results:

$$f(V_{opt}) = \frac{\dfrac{1}{m_r} + \displaystyle\sum_{j=1}^{r} \lambda_j}{\left(1 + \dfrac{t_a V}{t_m V_0} \right)} = \frac{\dfrac{1}{1.33} + 0.0}{\left(1 + \dfrac{1.27 \times V}{1.437 \times 40.0} \right)},$$

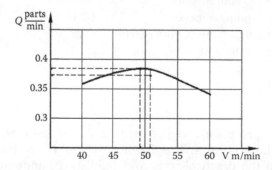

Figure 2.19 The productivity rate of machining the shaft by two cutters simultaneously versus the cutting speed.

$$g(V_{opt}) = \sum_{i=1}^{n}\left(\frac{1}{b_i}-1\right)\left(\frac{V}{C_i}\right)^{\frac{1}{b_i}} = \left(\frac{1}{0.15}-1\right)\left(\frac{V}{75}\right)^{\frac{1}{0.15}} + \left(\frac{1}{0.3}-1\right)\left(\frac{V}{400}\right)^{\frac{1}{0.3}}$$

Graphical solution is represented in Figure 2.20.

The intersection of two curves in Figure 2.20 gives the optimal cutting speed $V_{opt} = 49.0$ m/min. Substituting this speed into Eq. (2.16) and calculating yield the following result of maximum productivity:

$$Q_{max} = \frac{1}{\dfrac{t_m V_0}{V}+t_a} \times \frac{1}{1+m_r\left[\left(\dfrac{V}{C_1}\right)^{1/b_1} + \left(\dfrac{V}{C_2}\right)^{1/b_2}\right]}$$

$$= \frac{1}{\dfrac{1.437 \times 40.0}{49.0}+1.27} \times \frac{1}{1+1.33\times\left[\left(\dfrac{49.0}{75}\right)^{1/0.15} + \left(\dfrac{49.0}{400}\right)^{1/0.3}\right]}$$

$$= 0.379 \text{ work part/ min}$$

The theoretical optimal cutting speed for HSS cutter is 49 m/min that gives the maximum productivity $Q_{max} = 0.379$ part/min (Figure 2.19) can be executed by the speed of spindle rotation $n = \dfrac{1000V_0}{\pi d} = \dfrac{1000 \times 49.0}{\pi \times 25.0} =$ 623.887 rpm. The lathe machine has the stepped and discrete number of spindle's revolution. The nearest speed to the optimal of the spindle rotation for the lathe machine is $n = 650$ rpm, which corresponds to the cutting speed $V = 51.05$ m/min. This speed was used to validate the mathematical model of the productivity rate represented by Eq. (2.16). The confirmation

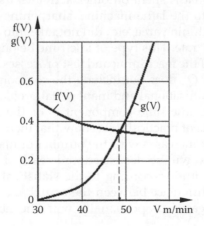

Figure 2.20 Graphical solution of the optimal cutting speed for the HSS cutter.

of some points of the equation means validation of mathematical model for productivity rate, i.e. the point of maximum productivity.

The theoretical productivity of the two cutter machining process with cutting speed $V = 51.05$ m/min is represented by the following result:

$$Q_{th} = \frac{1}{\dfrac{t_m V_0}{V} + t_a} \times \frac{1}{1 + m_r \left[\left(\dfrac{V}{C_1} \right)^{1/b_1} + \left(\dfrac{V}{C_2} \right)^{1/b_2} \right]}$$

$$= \frac{1}{\dfrac{1.437 \times 40.0}{51.05} + 1.27} \times \frac{1}{1 + 1.33 \times \left[\left(\dfrac{51.05}{75} \right)^{1/0.15} + \left(\dfrac{51.05}{400} \right)^{1/0.3} \right]}$$

$$= 0.377 \text{ work part/min}$$

where all other parameters are as specified earlier.

Actual productivity is calculated by the equation and substituting data presented in Table 2.4 into this equation yield the following result:

$$Q_{ac} = \frac{z}{\theta} = \frac{z}{T + \theta_i} = \frac{1}{t_m + t_a + (nm_r/z)} = \frac{1}{1.437 + 1.27 + (26 \times 1.33)/200}$$

$$= 0.347 \text{ work part/min}$$

where z is the number of the machined workpieces, θ is the time of the tests, T is the cycle time, $\theta_i = nm_r/z$ is the idle time due to cutters replacement referred to one product, n is the number of the cutters replacement and other parameters are as specified earlier.

The idle time that is spent on other activities like deliver the workpieces from stock to the lathe machine, sharpening of the HSS cutters, cleaning of the machining area, etc., did not take into account the calculation of productivity rate. This type of idle time belongs to organisational problems and out of the machining and test processes. The equation of the actual productivity Q_{ac} does not include the component of the idle time that is spent on organisational and managerial problems.

Analysis of the idle time component θ_i of the actual productivity equation and results of practical tests show that increasing the number of cutters replacement increases with the number of machined workpieces. However, their ratio will be decreased and reached to the defined constant average magnitude according to the statistical data. Hence, it can be predicted that difference between the actual and theoretical productivity will be decreased proportionally with the number of machined workpieces.

Conducted tests of machining the workpieces to cutters simultaneously and obtained data of the productivity rates enable to calculate the difference between theoretical and practical results as follows:

$$\delta = \frac{Q_{th} - Q_{ac}}{Q_{th}} = \frac{0.377 - 0.347}{0.377} \times 100\% = 7.95\%$$

The difference in results of actual and theoretical calculations is not greater in magnitude and accepts the mathematical model of the productivity rate as true (Figure 2.19). However, validation of the theoretical model is necessary to conduct deep and expanded tests according to the rules of mathematical statistics. It can be explained by the following reasons:

- The number of cutters replacement for the calculations of the average data for tool life and other parameters should be defined by recommendations of mathematical statistics.
- Accuracy of measurement of machining data, the cutter edge wearing level and service parameters were conducted manually and represented by average data.
- The Taylor equation of the tool life is empirical and practically the tool life of the cutters demonstrates quite a large deviation.

These restrictions in the tests of the machining process can give sensitive deviation in calculations of the cutters and machining parameters. Nevertheless, the represented results of the short practical tests are quite reassurable and confirm that the mathematical model of the productivity rate for machining workpieces by simultaneously engaged cutters can be accepted for practical use in the manufacturing area.

Analysis of the results calculated by Eqs. (2.16) and (2.19) and represented in Table 2.5 shows the following. The productivity rate of optimal machining the work part by a group of tools at conditions of simultaneous action ($Q = 0.106$ part/min) is higher than the use of normative machining process ($Q = 0.96$ part/min) and higher than the use of optimal machining processes for single operations ($Q = 0.98$ part/min).

Optimisation of machining process for single tools gives less value of productivity rate. This process does not give minimisation of time losses due to reliability of tools by Eq. (2.16). Hence, optimisation of machining processes for the single operations does not give the maximum productivity rate in case of simultaneous actions of tools. Equation (2.19) enables to minimise simultaneously the machining time and time losses due to reliability of tools.

The result of the analysis represented by the mathematical models of a machine tool's productivity rate, for single- and multi-tool machining processes enable the manufacturers to calculate a machine tool's output. The mathematical models for the productivity rate of a machine tool with respect to changes in the cutting speed include basic parameters of machining processes, i.e. machining time, coefficient of change in the cutting speed, reliability attributes of mechanisms and number of cutters in a machine tool. The equations contain the failure rates of both primary mechanisms and tools. Based on the new equations, the productivity rate of single- and multi-tool processes for a machine tool are calculated as a function of the change in cutting speed. The new equations for the productivity rate make it possible to determine the optimal cutting speed of single- and multi-tooling processes in simultaneous action, which in turn can give the maximum productivity rate of a machine tool. The optimal change in cutting speed depends on the following factors: cutter's material, machining and auxiliary times, mean repair time and failure rates of all machine tool components. The equations enable the prediction of more authentic results and can be used in preparing economically effective multi-tool manufacturing processes.

The desire to increase the productivity rate of machine tools has led to strategies that involve intensifying the manufacturing processes and finding mathematical models for the maximum productivity rate of machine tools. The optimisation of a machining process using the criterion of maximum productivity rate is therefore crucial. Finding mathematical models for productivity rate presents a holistic picture of the machine tool work, the parameters of technology and the reliability of machining processes. The new equations include basic parameters of machining processes, machining time, the cutting speed, reliability attributes of mechanisms and cutters of a machine tool. The equations of the productivity rate with single- and multi-tool machining processes of simultaneous action enables to find the optimal cutting speed for each cutter. The maximum productivity rate obtained by the minimisation of the machining time, optimisation of the failure rates of the cutters and optimisation of their cutting speeds for multi-tooling processes. Optimisation of the cutting speeds for each cutter for single operations does not lead to maximum productivity in case of the simultaneous engagement in machining process. Formulated equations for productivity rate enable prediction more authentic results and can be used in preparing economically effective multi-tool manufacturing processes of work parts.

2.4.2   The productivity rate of the machine tool with the single machining process

Equation (2.16) for the productivity rate of multi-tool processing of the work part enables simplification to derive the equation for the productivity rate of machining the work part by the single cutter. Modified Eq. (2.16) thus gives the following equation:

$$Q = \frac{1}{\frac{t_{m.n}V_n}{V} + t_a} \times \frac{1}{1 + m_r\left[\left(\frac{V}{C}\right)^{1/b} + \sum_{i=1}^{r}\lambda_j\right]} \tag{2.20}$$

where all parameters are as specified earlier.

The optimal cutting speed for the single-tool machining process by criterion of the maximal productivity rate is obtained by solving the first derivative of Eq. (2.20) with respect to the cutting speed and setting it to zero. Then, the first derivative yields the following differential equation:

$$\frac{dQ}{dV} = \frac{d\left\{\frac{1}{\frac{t_{m.n}V_n}{V} + t_a} \times \frac{1}{1 + m_r\left[\left(\frac{V}{C}\right)^{1/b} + \sum_{i=1}^{r}\lambda_j\right]}\right\}}{dV} = 0$$

giving rise to the following

$$\frac{t_{m.n}V_n}{V^2\left(\frac{t_{m.n}V_n}{V} + t_a\right)^2} \times \frac{1}{1 + m_r\left[\left(\frac{V}{C}\right)^{1/b} + \sum_{j=1}^{r}\lambda_j\right]} - \frac{1}{\frac{t_{m.n}V_n}{V} + t_a}$$

$$\times \frac{m_r\left(\frac{1}{b}\right)\left(\frac{V}{C}\right)^{(1/b)-1}}{\left\{1 + m_r\left[\left(\frac{V}{C}\right)^{1/b} + \sum_{j=1}^{r}\lambda_j\right]\right\}^2} = 0 \tag{2.21}$$

Simplification and transformation of Eq. (2.21) yield the following equation:

$$\frac{1}{m_r} + \sum_{j=1}^{r}\lambda_j + \left(\frac{V}{C}\right)^{\frac{1}{b}} = \frac{1}{b}\left(\frac{V}{C}\right)^{\frac{1}{b}}\left(1 + \frac{t_a V}{t_{m.n}V_n}\right) \tag{2.22}$$

where V is the optimal cutting speed and other parameters are as specified earlier.

Equation (2.22) is transcendental, for which the graphical solution is similar to Eq. (2.18) with two equations.

$$f_i(V_{opt}) = \frac{1}{m_r} + \sum_{j=1}^{r} \lambda_j + \left(\frac{V}{C}\right)^{\frac{1}{b}}, \quad g(V_{opt}) = \frac{1}{b}\left(\frac{V}{C}\right)^{\frac{1}{b}}\left(1 + \frac{t_a V}{t_{m.n} V_n}\right) \quad (2.23)$$

Analysis of the change of argument V in increasing these functions $f(V_{opt})$ and $g(V_{opt})$ with given data demonstrates the same alterations as described for Eq. (2.19). The intersection of functions $f(V_{opt})$ and $g(V_{opt})$ gives the solution of the optimal cutting speed for the machining process by criterion of the maximal productivity rate.

Naturally, this solution can be compared with the optimal cutting speed for the machining process by criterion of the minimal cost, for which mathematical models is well described in literature with many variations and nuances in computing. One of the common equation for the optimal cutting speed is represented by the following expression [7]:

$$V_{opt} = \frac{C(L_m + B_m)^b}{\{[(1/b) - 1][C_s + T_c(L_m + B_m) + T_g(L_g + B_g) + D_c]\}^b}, \quad (2.24)$$

where C is the empirical constant resulting from regression analysis and field studies; L_m is the labour cost of production operator per hour; B_m is the burden rate, or overhead charge, of the machine, including depreciation, maintenance, indirect labour and the like; b is the empirical constant that generally depends on the cutter tool material; T_c is the time required to change the tool; T_g is the time required to grind the tool; L_g is the labour cost of tool-grinder operator per hour; B_g is the burden rate of tool grinder per hour and D_c is the depreciation of the tool in dollars per grind.

Equation of the cost of the product is represented by the following equation:

$$C_p = (L_m + B_m)(T_m + T_l) + C_s + T_m\left(\frac{V}{C}\right)^{\frac{1}{b}}[T_c(L_m + B_m) + T_g(L_g + B_g) + D_c]$$

$$(2.25)$$

where T_m is the machining time per piece; C_s is the cost of setting up for machining, such as mounting the cutter and fixtures and preparing the machine for the particular operation; T_l is the time involved in loading and unloading the part, changing speeds, changing feed rates and so on; T_l is the time involved in loading and unloading the part, changing speeds, changing feed rates and so on; other parameters are as specified earlier.

Obviously, two different criterions for the optimal cutting speed (Eqs. 2.23 and 2.24) give different values. Which optimal cutting speed and its criterion are necessary to choose depend on the marketing environment and solution of the marketing managers. As a rough guide is proposed the following approaches. When a manufacturer fabricates a product without concurrency and the market consumes these products, in such case, it is preferably to choose the optimal cutting speed defined by criterion of the maximal productivity rate. In case, when a market demonstrates similar products, the optimal cutting speed is defined by criterion of the minimal cost that should give the lowest price for a product and enables for products to be competitive.

A working example

The single-tool machining process is represented by simple turning operation of the shaft. The turning cutter on the lathe tool machines the carbon steel work part. The machining and economic parameters and properties of the turning operation are represented in Table 2.6 [1].

The optimal cutting speed for turning operation should be defined using two criteria, namely the maximal productivity rate and minimal cost.

Solution:

1. The optimal cutting speeds using the criterion of maximal productivity rate for turning processes is defined by Eq. (2.23).

Table 2.6 Machining and tool's normative parameters

Tool	Turning cutter
Tool material	High speed steel
Machining time, $t_{m.n} = T_m$ (min/work part)	4.0
Auxiliary time, $t_a = T_l$ (min/work part)	0.3
Cutter's constant C	75
b	0.15
Tool life, T (min)	50
Normative cutting speed, V_n (m/min)	40
Failure rate of machine units, $\sum_{j=1}^{r} \lambda_j$	6×10^{-7}
Mean repair time, $m_r = T_c$ (min)	3.0
Depreciation of the tool in dollars per grind, D_c ($/min)	1.142
Labour cost of production operator per hour, L_m ($/min)	0.071
Burden rate of the machine, including depreciation, maintenance, indirect labour and the like, B_m ($/min)	0.238
Labour cost of tool-grinder operator per hour, L_g ($/min)	0.094
Burden rate of tool grinder per hour, B_g ($/min)	0.285
Cost of setting up for machining 50 work parts, C_s ($)	0.4
Time required to grind the tool, T_g (min)	3.0
Time involved in loading and unloading the work part, T_l (min)	0.3

Substituting the given parameters (Table 2.6) into defined equations and calculating yield the optimal cutting speeds.

The optimal cutting speed of the turning cutter that gives the maximal productivity rate is as follows:

$$f_i(V_{opt}) = \frac{1}{m_r} + \sum_{j=1}^{r} \lambda_j + \left(\frac{V}{C}\right)^{\frac{1}{b}} = \frac{1}{3.0} + 6 \times 10^{-7} + \left(\frac{V}{75}\right)^{\frac{1}{0.15}}$$

$$g(V_{opt}) = \frac{1}{b}\left(\frac{V}{C}\right)^{\frac{1}{b}}\left(1 + \frac{t_a V}{t_{m.n} V_n}\right) = \frac{1}{0.15}\left(\frac{V}{75}\right)^{\frac{1}{0.15}}\left(1 + \frac{0.3 \times V}{4.0 \times 40}\right)$$

Graphical solution of the optimal cutting speed for the turning operations is represented in Figure 2.21.

Substituting the optimal cutting speed $V_{opt} = 48$ m/min (Figure 2.21) and other parameters from Table 2.6 into Eq. (2.16) and computing yield the maximal productivity rate for the turning operation.

$$Q_{max} = \frac{1}{\frac{4.0 \times 40}{48} + 0.3} \times \frac{1}{1 + 3\left[\left(\frac{48}{75}\right)^{\frac{1}{0.15}} + 6 \times 10^{-7}\right]}$$

$$= 0.238 \text{ product/min} = 14.321 \text{ product/h}$$

2. Equations (2.24) and (2.25) define the optimal cutting speed by criterion of minimal cost of turning operation and the cost the product. Substituting the given parameters (Table 2.6) into defined equations and calculating yield the following results.

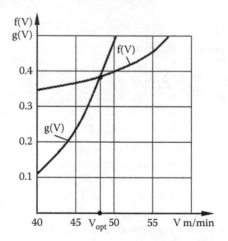

Figure 2.21 Graphical solution of the optimal cutting speed for turning process by criterion of the maximal productivity rate.

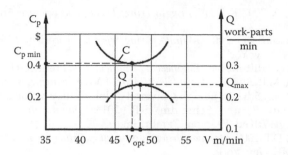

Figure 2.22 Productivity rate and minimal cost of the turning process versus the cutting speed.

$$V_{opt} = \frac{75(0.071+0.238)^{0.15}}{\{[(1/0.15)-1][(3/60)(0.071+0.238)+(3.0/60)(0.094+0.285)+1.142]\}^{0.15}}$$

$$= 47.318 \, m/min$$

$$C_{p.min} = (0.071+0.238)(4+0.3)/60+0.4$$

$$+ \frac{4}{60}\left(\frac{48}{75}\right)^{\frac{1}{0.15}}[(3/60)(0.071+0.238)+(3/60)(0.094+0.285)$$

$$+1.142] = 0.426 \, \$$$

Obtained results represented in Figure 2.22 demonstrates two diagrams of the optimal cutting speeds for the maximal productivity rate and for the minimal machining cost of the turning process versus the change in the cutting speed. The diagrams focused on results that are most important information.

Analysis of the diagrams in Figure 2.22 demonstrates the optimal cutting speeds are different for the criterions of the maximal productivity rate and minimal machining cost. This is natural result due to difference in origin approaches.

The optimal cutting speed by criterion minimum cost is computed for machining 50 work parts. With increasing the number of machining products, the cost of setting up and other related parameters is decreased. Then, the difference in optimal cutting speeds computed by the maximal productivity rate and minimum cost is decreased. This result confirms result of practical economics of manufacturing processes.

Bibliography

1. Badiru, A.B., and Omitaomu, O.A. 2011. *Handbook of Industrial Engineering Equations. Formulas, and Calculations*. Taylor & Francis. New York.
2. Benhabib, B. 2005. *Manufacturing: Design, Production, Automation, and Integration*. 1st ed. Taylor & Francis. New York.

3. Birolini, A. 2007. *Reliability Engineering: Theory and Practice.* 5th ed. Springer. New York.
4. Chryssolouris, G. 2006. *Manufacturing Systems: Theory and Practice.* 2nd ed. Springer. New York.
5. Games, G. 2010. *Modern Engineering Mathematics.* Prentice Hall. London.
6. Groover, M.P. 2013. *Fundamentals of Modern Manufacturing: Materials, Processes, and Systems.* 5th ed. (Lehigh University). John Wiley & Sons. Hoboken, NJ.
7. Kalpakjian, S., and Schmidm, S.R. 2013. *Manufacturing Engineering & Technology.* 7th ed. Pearson. Cambridge.
8. Koenig, D.T. 2007. *Manufacturing Engineering: Principles for Optimization.* 3rd ed. ASME. New York.
9. Mukherjee, I., and Ray, P.K. 2006. A review of optimization techniques in metal cutting processes. *Computer and Industrial Engineering* 50(1). pp. 15–34.
10. O'Connor, P.D.T., and Kleyner, A. 2012. *Practical Reliability Engineering.* 5th ed. John Willey & Sons. West Sussex.
11. Rao, R.V. 2013. *Manufacturing Technology - Vols. 1; 2.* 3rd ed. McGraw Hill Education. New York.
12. Rao, R.V. 2011. *Advanced Modeling and Optimization of Manufacturing Processes.* 1st ed. Springer Series in Advanced Manufacturing. Springer-Verlag. New York.
13. Schey, J. 2012. *Introduction to Manufacturing Processes.* 3rd ed. McGraw-Hill Education. New York.
14. Shaumian, G.A. 1973. *Complex Automation of Production Processes.* Mashinostroenie. Moscow.
15. Usubamatov, R., Zain, Z.M., Sin, T.C., and Kapaeva, S. 2016. Optimization of multi-tool machining processes with simultaneous action. *The International Journal of Advanced Manufacturing Technology.* DOI: 10.1007/s00170-015-6920-x.
16. Usubamatov, R., Ismail, K.A., and Shah, J.M. 2012. Mathematical models for productivity and availability of automated lines. *International Journal of Advanced Manufacturing Technology.* DOI: 10.1007/s00170-012-4305-y.
17. Volchkevich, L. 2005. *Automation of Production Processes.* Mashinostroenie. Moscow.

chapter three

Manufacturing systems of parallel arrangement and productivity rate

Industrial processes in different areas, such as pressworks, assembling, liquid filling for different types of container and other similar technological processes, involve very short operation cycle time and defined productivity rate. This type of process is implemented on one machine tool. Demand to implement the productivity output leads to install several parallel machine tools or workstations with same technological processes. Machine tools and workstations arranged in parallel makeup create the manufacturing system of parallel structures with linear and rotor-type automatic machines. The latter one finds wide applications in various branches of the industries with technologies that affect the surface and volume of products. Mathematical models for the productivity rate of the linear and rotor-type machines of parallel structures are different and derived from certain analytical approaches that consider technological parameters, principle work of a rotor-type machine, reliability of machine units and a number of parallel workstations.

The comparative analysis of the productivity rates for manufacturing machines of parallel arrangements is implemented, irrelative to the specific area of their applications in industries. The highest productivity rate gives the manufacturing system of parallel structure with independent workstations and the lowest rate gives the rotor-type machine. Nevertheless, in most cases, manufacturers prefer to use the rotor-type machine for machining products due to economic reasons in spite of its lowest productivity rates.

3.1 Introduction

Industrial processes in different areas, such as pressworks (stamping, drawing, punching, trimming, coining), assembling, liquid filling for different types of container (bottles, cans) and the other similar technological processes, involve very short operation cycle times. Practically, the short-type technological processes do not need further decomposition of the technological processes with the aim of increasing the productivity

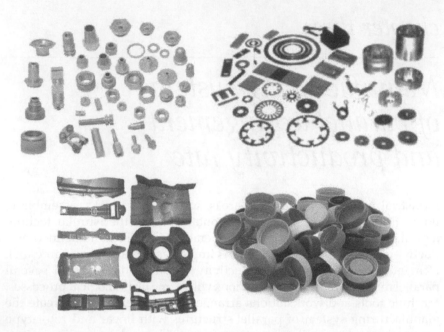

Figure 3.1 Typical work parts from different materials with short technological processes.

of industrial machines. These types of processes are implemented on one single workstation or a machine tool. The typical parts of short technological processes are presented in Figure 3.1.

In manufacturing industries, there are numerous types of products with simple design that are machined by single machine tools. However, in many cases, these machine tools cannot carry out the company's output program. Demand to implement productivity output leads to install several parallel machine tools or workstations with the same technological processes. This type of parallel machine tools arrangement is presented as the manufacturing system of parallel structures. In industries, machines and automated lines of parallel arrangement different in layout and designs are represented by the following solutions:

- Manufacturing lines of parallel arrangement with independent work of machine tools or workstations.
- Automated lines of parallel structure.
- Automatic machine of circular design (rotor-type machine) and parallel structure.

These three types of manufacturing systems have different properties and mathematical models for the productivity rate. The following analysis of properties and area of application for different types of manufacturing systems of parallel arrangements enables for developing reliable mathematical models to calculate the manufacturing system's productivity rate. The productivity rate for manufacturing system of parallel arrangement is increased proportionally to the number of parallel workstations. Increasing the number of workstations in the manufacturing systems leads to change in reliability and productivity indices of a complex system. Solving the problem of the structure for manufacturing systems by the criterion of maximal productivity and efficiency is a crucial issue. Detailed considerations of mathematical models for each type of parallel manufacturing systems and automated lines are presented below.

3.2 Productivity rate of manufacturing systems with parallel arrangement and independent work of workstations

Machine tools and workstations arranged in parallel makeup create the manufacturing system of parallel structures, which is described by the following properties:

- The parallel technological process for machining the simple parts is implemented on several machine tools according to the output requirements.
- The parallel arrangement of the machine tools for machining the parts is presented as the manufacturing system with independent work of each machine tool.
- The parallel arrangement of the independent machine tool occupies a large manufacturing area.
- The work parts are feeding to each machine tool manually or by simple means of mechanisation and automation (slider, roller, trolley, feeder, etc.)
- The failure of any machine tool in the manufacturing system of parallel arrangement does not lead to a stop of other machine tools and entire system.
- The machining time on each machine tool is same as other machine tools in the manufacturing system.
- Productivity rate for the manufacturing system of parallel arrangement is represented by a single machine tool in the system and multiplied by the number of parallel machine tools.

Figure 3.2 presents the views of manufacturing systems or lines of parallel arrangement in different industries.

(a)   (b)

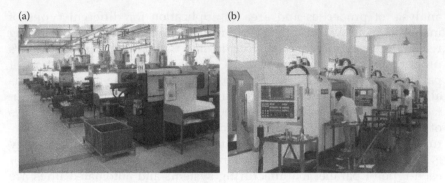

Figure 3.2 Manufacturing lines of parallel arrangement.

A symbolic picture of the manufacturing line of parallel arrangement with independent work of p machine tools is presented in Figure 3.3.

The properties of the manufacturing line of parallel arrangement represented in the earlier figure enable for writing the mathematical model of the productivity rate based on Eq. (2.10) for a single machine tool. Hence, the productivity rate of manufacturing line of parallel arrangement with independent work of machine tools is represented by the following equation:

$$Q = \frac{p}{t_m + t_a} \times \frac{1}{1 + m_r \sum_{i=1}^{g} \lambda_i} = \frac{p}{t_m + t_a} \times \frac{1}{1 + m_r \lambda_s} \tag{3.1}$$

where p is the number of parallel products machined equal to the number of parallel machine tools or workstations ($p = p_s$) arranged in the manufacturing line; λ_s is the failure rate of a workstation. All other parameters are as specified earlier.

Component of Eq. (3.1) is the availability $A = 1/(1 + m_r \lambda_s)$ that is considered only for one workstation, because the other parallel workstations are independent, and there are no mechanical or other type connections between them. Other parallel machines or workstations are implementing same process and are presented in Eq. (3.1) as the number p products machined by the p_s parallel workstations. Analysis of Eq. (3.1) enables for depicting the change in productivity rate of the manufacturing line of parallel arrangement with increase in the number of parallel machines in the line. This change is represented in Figure 3.4, which demonstrates a

1   2   3   ...   i   ...   p

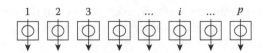

Figure 3.3 Symbolic presentation of the manufacturing line of parallel arrangement.

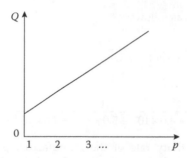

Figure 3.4 The productivity rate of a manufacturing line of parallel arrangement versus the number of independent parallel workstations.

linear increase in the productivity rate for a manufacturing line of parallel structure versus the number of machine tools or workstations.

Figure 3.4 depicts that the increase in the productivity rate for the manufacturing line of parallel arrangement increases with the number of parallel machines with independent work. Random failure on one machine does not lead to a stop of other machines.

A working example

Manufacturing line of parallel arrangement with independent machine tools or workstations is designed with the technical and timing data referred to one product and is shown in Table 3.1. The cyclic and total productivity rates and availability for the given data of a manufacturing line of parallel arrangement are calculated.

Solution: Substituting the initial data of Table 3.1 into Eq. (3.1) and calculating lead to the following results.
    The productivity rate per cycle is as follows:

$$Q = \frac{1}{t_m + t_a} = \frac{1}{6.0 + 0.3} = 0.158 \text{ work part/min}$$

Table 3.1 Technical data of a manufacturing system of parallel arrangement

Title		Data
Machining time, t_{mo} (min)		6.0
Auxiliary time, t_a (min)		0.3
Failure rate of the unit, λ_i (per min)	Cutter, λ_c	0.0125
	Spindle, λ_s	4×10^{-8}
	Support, λ_{sp}	5×10^{-9}
	Gearing, λ_g	4×10^{-8}
	Control system, λ_{cs}	6×10^{-9}
Mean repair time, m_r (min)		3.0
Number of parallel stations, p_s		10

The availability is as follows:

$$A = \cfrac{1}{1 + m_r \sum_{i=1}^{g} \lambda_i} = \cfrac{1}{1 + m_r(\lambda_c + \lambda_s + \lambda_{sp} + \lambda_g + \lambda_{cs})}$$

$$= \frac{1}{1 + 3.0(0.0125 + 4.0 \times 10^{-8} + 5.0 \times 10^{-9} + 4.0 \times 10^{-8} + 6.0 \times 10^{-9})} = 0.96$$

The total productivity rate of the manufacturing line of parallel arrangement with independent action of workstations is as follows:

$$Q = \frac{p}{t_m + t_a} \times \cfrac{1}{1 + m_r \sum_{i=1}^{g} \lambda_i} = 10 \times 0.158 \times 0.96 = 1.52 \text{ work part/min}$$

The calculations demonstrate that the manufacturing system of parallel arrangement with independent work of parallel machines or workstations does not have a limit of productivity rate and depends on the number of machines. The team of technicians serve the manufacturing system and any random failure of machines is repaired for the short time. This is the reason that Eq. (3.1) is acceptable and considered that all machines of manufacturing systems are in workable conditions.

3.3 Productivity rate of automated lines of parallel structure and linear arrangement

The demand to increase machine productivity rate for the machine tools or workstations with short operation cycle time leads to the design of parallel workstations on one work bed that forms an automated line of parallel structure. This type of automated line is designed on the base of identical machine tools (workstations) in parallel structure, which is described by the following properties:

- The parallel disposition of the identical workstations for machining the work parts represented the automated line parallel structure that is designed on one bed with one control system and hard linking of all workstations and mechanisms.
- The automated line of parallel structure occupies less manufacturing area than a manufacturing line with independent machine tools and uses less volume of metals, materials, control systems and other design components in construction.
- The failure of any workstation, mechanism or control units leads to a stop of the entire automated line of parallel structure that results in a decrease in the productivity rate.

- Each workstation in the automated line of parallel structure is furnished by automatic work-part feeders.
- Productivity rate for the automated line of parallel structure is presented as a product of the productivity of one workstation, the number of parallel workstations and availability that modified according the increased failure of an automated line.
- All workstations in an automated line of parallel structure start machining process at one time and finish at one time, i.e. the cycle time of machining process is identical at all parallel workstations.

Figure 3.5 represents the views of automated lines of parallel structure in different industries.

A symbolic picture of the automated line of parallel structure and linear arrangement with p workstations is represented in Figure 3.6.

The diagram of the work of an automated line of parallel structure is represented in Figure 3.7.

Diagram depicts the starting and ending of machining process at all workstations at an automated line of parallel structure. The properties of

Figure 3.5 Automated lines of parallel structure in (a) textile industry, (b) manufacturing industry, (c) food industry, (d) electronic industry.

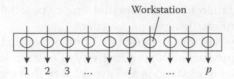

Figure 3.6 A symbolic presentation of automated lines of parallel structure and linear arrangement.

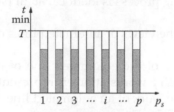

Figure 3.7 Diagram of a work for the manufacturing and automated line of parallel structure and linear arrangement with p_s parallel workstations.

the automated line of parallel action presented earlier enable for presenting the mathematical model for productivity rate based on Eq. (3.1) for a manufacturing line. Hence, the productivity rate of automated line of parallel structure is represented by the following equation [10,14]:

$$Q = \frac{p}{t_m + t_a} \times \frac{1}{1 + m_r[p_s(\lambda_s + \lambda_f) + \lambda_{cs}]} \tag{3.2}$$

where failure rate of an automated line of parallel structure is increased p_s times according to the number of parallel workstations, units and mechanisms, λ_s is the failure rate of one workstation, λ_f is the failure rate of the work-part feeder for one workstation, λ_{cs} is the failure rate of common control system. All other parameters are as specified earlier.

Component of Eq. (3.2) is the availability $A = 1/(1 + m_r[p_s(\lambda_s + \lambda_f) + \lambda_{cs}])$ and includes p_s workstations, because the hard linking of all workstations in the automated line leads to a proportional increase in p_s times the failure rates of all other parallel workstations and decrease in the value of the availability. Productivity rate of the automated line of parallel structure is calculated for one station. Other parallel machines are implementing the same process and presented in Eq. (3.2) as the number p work parts machined by the p_s parallel workstations. Equation (3.2) demonstrates that the increase in the number of parallel workstations p_s leads to an increase in the productivity rate per cycle, and at the same time, the value of the availability for the automated line decreases.

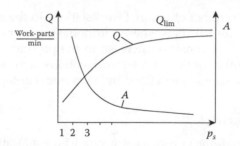

Figure 3.8 Change in productivity rate per cycle and availability of automated line of parallel structure versus change in the number of parallel workstations.

Analysis of Eq. (3.2) demonstrates there is one component in denominator, $m_r p_s (\lambda_s + \lambda_f)$, whose value is increasing with increase in the numbers of workstations p_s. Other components are not changing and have constant values, t_m, t_a and $m_r \lambda_{cs}$. Figure 3.8 depicts the change in variable components of the Eq. (3.2) versus the change in the number of parallel workstations p_s.

Preliminary analysis of Eq. (3.2) demonstrates that there is some limit of productivity rate for the automated line of parallel structure. The maximal productivity rate Q_{max} is defined by solving a mathematical limit of Eq. (3.2)

$$\underset{p \to \infty}{\text{Lim}} Q = \underset{p \to \infty}{\text{Lim}} \left(\frac{p}{t_m + t_a} \times \frac{1}{1 + m_r p_s (\lambda_s + \lambda_f)} \right) = \frac{1}{(t_m + t_a)(\lambda_s + + \lambda_f) m_r} \quad (3.3)$$

The maximal productivity for the automated line of parallel structure and linear arrangement depends on the cycle time and reliability parameters of one workstation.

Figure 3.9 depicted by Eq. (3.3) demonstrates that the increase in productivity rate for the automated line of parallel structure with increase in the number of parallel workstations is asymptotically reached to defined limit.

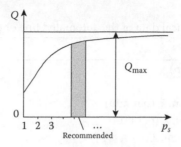

Figure 3.9 The productivity rate of an automated line of parallel structure versus the number of parallel workstations.

The optimal number of parallel workstations in the automated line is recommended to find at zone, where additional numbers of parallel workstation do not give a sensitive increase in the productivity rate. Finally, the maximal number of the parallel workstations at an automated line of parallel structure should be defined by economical criterion.

A working example 1

The automated line of parallel structure with workstations of hard connecting is designed with the technical and timing data referred to one product and is shown in Table 3.2. The cyclic, total and maximum productivity rates and availability for the given data of the automated line of parallel structure are calculated.

Solution: Substituting the initial data of Table 3.2 into Eqs. (3.2) and (3.3) calculating lead to the following results.

The productivity rate per cycle is as follows:

$$Q = \frac{1}{t_m + t_a} = \frac{1}{6.0 + 0.3} = 0.158 \text{ work part/min}$$

The availability is as follows:

$$A = \frac{1}{1 + m_r[p_s(\lambda_s + \lambda_f) + \lambda_{cs}]} = \frac{1}{1 + 3.0[10 \times (0.0125 + 3.0 \times 10^{-10}) + 6.0 \times 10^{-8})}$$

$$= 0.72$$

The total productivity of the automated line of parallel structure is as follows:

$$Q = \frac{p}{t_m + t_a} \times \frac{1}{1 + m_r[p_s(\lambda_s + \lambda_f) + \lambda_{cs}]} = 10 \times 0.158 \times 0.72$$

$$= 1.15 \text{ work part/min}$$

The maximal productivity rate of the automated line of parallel structure is as follows:

Table 3.2 Technical data of the automated line of parallel structure

Title	Data
Machining time, t_{mo} (min)	6.0
Auxiliary time, t_a (min)	0.3
Failure rate of the station, λ_s (per min)	0.0125
Failure rate of the feeder, λ_f (per min)	3×10^{-10}
Failure rate control system, λ_{cs}	6×10^{-8}
Mean repair time, m_r (min)	3.0
Number of parallel machine tools, p_s	10

$$Q_{max} = \frac{1}{(t_m + t_a)m_r(\lambda_s + + \lambda_f)} = \frac{1}{(6.0 + 0.3) \times 3.0 \times (0.0125 + 3.0 \times 10^{-10})}$$

$$= 4.23 \text{ work part/min}$$

Results of computing are represented in Figure 3.10, which demonstrates that the productivity rate increases significantly for a small number of parallel workstations and asymptotically reaches the limit of $Q_{max} = 4.23$ work part/min. The optimal number of parallel stations should be defined by economic approaches for the automated line with parallel stations.

A working example 2

The automated line of parallel structures (Figure 4.6) can be designed with variants, for which the technical data are presented in Table 3.3. Substituting the given data into Eq. (3.2) and transformation yield the results for the availability of the automated line of parallel structure and its productivity rate versus the failure rate. Figure 3.11 demonstrates the change in the availability and productivity rate for

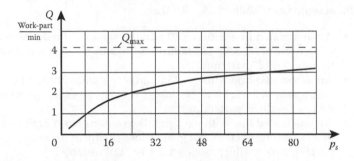

Figure 3.10 Productivity rate of an automated line with parallel structure versus the number of parallel workstations.

Table 3.3 Technical data for an automated line of parallel structure

Title	Data
Total machining time, t_{mo} (min)	1.0
Auxiliary time, t_a (min)	0.1
Number of parallel workstations, p_s	5, ..., 15
Reliability attributes for an automated line	
Average failure rate of the workstation, λ_s (fail/min)	$\lambda \times 10^{-3}$
Failure rate of the control system, λ_{cs}	$\lambda \times 10^{-8}$
Failure rate of the transport system, λ_{tr}	$\lambda \times 10^{-10}$
Mean repair time, m_r	3.0 min

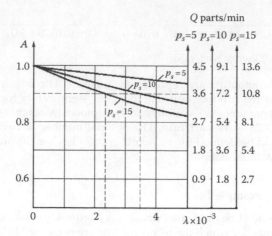

Figure 3.11 Productivity rate and availability of the automated line with p_s parallel workstations versus the increase in the failure rate λ of the automated line.

the automated lines of parallel structure with a different number of stations p_s versus the change in the failure rates.
If the accepted availability is $A = 0.9$, then

- An automated line with $p = 10$ parallel stations should have a failure rate of $\lambda \leq 3.5 \times 10^{-3}$ failures/min, and gives a productivity rate $Q = 7.27$ part/min.
- An automated line with $p = 15$ parallel stations should have a failure rate of $\lambda \leq 2.3 \times 10^{-3}$ failures/min and gives a productivity rate $Q = 10.88$ part/min.
- An automated line with $p = 5$ parallel stations has a high level of availability $A = 0.93$, but gives a low productivity rate $Q = 4$ part/min with failure rate $\lambda = 5.0 \times 10^{-3}$ failures/min.

The diagrams demonstrate that increase in the failure rate of the automated line of parallel structure decreases the value of availability in the automated line. Increase in the number of parallel workstations in the automated line reflects on the values of availability and productivity rate in the automated line: increase in the number of workstations leads to a decrease in the value of availability and increase in the value of productivity rate. However, this dependency does not give extreme solution.

3.4   An automatic machine of parallel structure and circular design (rotor-type machine)

A rotor-type automatic machine belongs to the machine tool of parallel structure and action. Such automatic machines find wide application in various branches of the industries. Generally, the application of a rotor-type automatic machine is especially effective for machining of work parts or products by technologies that affect the surface and volume of products. Usually, in manufacturing area, these technologies are conducted with short operating cycles and with high frequency of loading and unloading of work parts. Rotary principle of a work can be used to produce work parts in one zone, which makes it possible build the automatic machine into an automated rotor-type line. Differences in designs of linear and rotor-type machines are in construction. An automated line of parallel structure is designed with the embedded work-part feeders, whose numbers are equal to the numbers of parallel workstations. One feeder that supplies the work parts to all parallel stations in series as the workstations approach to load zone furnishes the rotor-type machines. However, this specific design of a machine with one feeder leads to construct workstations with the ability to move in sectors of loading the work parts. Such requirement created circular arrangements of the workstations in one unit with the ability to rotate and deliver workstations to the work-part feeder in loading sector. This construction is formed as a rotor-type automatic machine with identical parallel workstations embedded in one rotor-type mechanism. Thus, a rotor-type machine is characterised by the fact that the workstations for machining of the work parts are moved continuously in a circle. These parallel workstations repeatedly and discretely affect the work parts, which are moved in a circle with the same transport speed and having contact with the machining tools one at a defined time. The machining process at each workstations is shifted at the time. Typical views of the rotor-type machines of parallel structure are represented in Figure 3.12.

For different industries, the rotor-type of automatic machines are designed with different number of workstations. In manufacturing industry, the rotor-type machines usually include 6–8 parallel workstations. The electric, electronic and food processing industries exploit

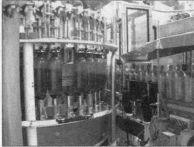

Figure 3.12 Rotor-type automatic machine of parallel structure for filling the bottles by liquids.

rotor-type machines that contain 20–40 parallel workstations. The number of parallel workstations at one rotor machine depends on the severity of machining process and size of a product to be machined. The advantage of rotary automatic machines is a continuous process that enables avoiding the inertial loads, which makes it possible to conduct short-time machining process with a high productivity rate. Furthermore, all the working zones are stationary and the loading/unloading mechanism is one for the rotary station. Workstation's identity properties enable the easiness and simplicity in creating automatic rotor-type lines.

Rotor-type automatic machine is designed on a base of identical workstations with parallel structure, which is described by the following properties:

- Automatic machines with circular arrangement of workstations are more compact and more convenient from the point of view of maintenance.
- The parallel structure of the identical workstations for machining the work parts is represented in one rotor mechanism with one control system and hard linking of all parallel workstations and mechanisms.
- Machining process on each workstation is implemented by a tool acting on the work part during circular motions. A set of parallel tools and work parts of workstations have circular motion provided by the rotor mechanism.
- Implementation of technological processes on each workstation at a rotor-type machine shifted on a defined time.
- The failure of any workstation, mechanism or control units leads to entire stop of a rotor-type automatic machine and increase in downtime.
- One automatic work-part feeder furnishes a rotor-type automated machine.
- Productivity rate for a rotor-type machine of parallel structure is represented as a product of productivity of the single workstation, the number of parallel workstations and availability that modified according to the shifted machining time and increased failures.

Figure 3.13 represents a sketch of rotor-type automatic machine that enables easy tracing of its work process. In the periphery of the working rotor, which has a continuous transport motion, disposed parallel workstations with tools are available for implementing the assigned operations. During the motion of working rotor by means of the transport mechanism, a feeder is loading the work parts consequently into the workstations, which are equipped with tools [12].

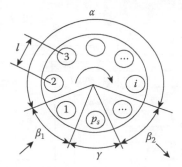

Figure 3.13 The sectors of activity of a rotor-type automatic machine of parallel structure with p_s parallel workstations.

The work of the rotor-type machine of parallel structure is represented by defined sectors of the workstations activity by the following components:

- In sector β_1, the feeder loads the work part to the machining area of the workstation, the work part is clamped and the tool starts approach to the work part rapidly.
- In sector α, the tool performs the technological motions (stamping, drawing, filling, assembling, etc.).
- In sector β_2, the tool is withdrawn from the machining area of a workstation, and the work part is released and moved out from the rotor machine.
- In sector γ, the tool is inspected, cleaned or replaced. This sector is free from machining process.

In other words, the sector α implements the machining process (t_m), whereas sectors β_1 and β_2 implements auxiliary motions (t_a). The diagram of a work for a rotor-type automatic machine of parallel structure and action is represented in Figure 3.14.

The diagram depicts the starting, displacement and ending of machining process at each parallel workstation of the rotor-type automated line. This diagram enables for formulating the productivity rate of the rotor-type automatic machine.

3.4.1   Productivity rate of rotor-type automatic machines

The fundamental basis of mathematical modelling for the productivity rate of the automatic machines of different structure is identical, because all methods of analysis and synthesis presented earlier are identical and can be applied to any types of manufacturing machines and system

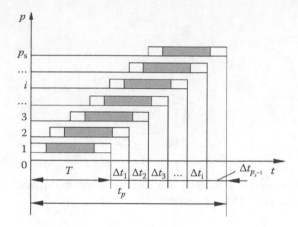

Figure 3.14 Diagram of a work for a rotor-type automatic machine of parallel structure with p_s parallel workstations.

designs. The rotary automated machine with simple technological processes can have variable structures based on the number of parallel workstations. The structure of such rotary machine depends on the duration of a cycle of technological process for processing of a work part, reliability of the rotary machine mechanisms and demanded productivity rate.

Mathematical models for the productivity rate of the rotor-type machines are complex and are currently derived on certain analytical approaches [12]. For the analysis of the productivity rate of the rotor-type automatic machines of parallel structure, it is necessary to consider the influence of all factors involved in the manufacturing process, such as technological parameters of machining process, principle of working of the rotor-type machine, reliability of machine units and number of parallel workstations.

Analysis of the work of a rotary automatic machine at the time is conducted in a common form. The rotary machine includes p_s parallel workstations and the same quantity of work parts involved in machining process. A diagram in Figure 3.14 represents the displaced time of machining process on each workstation in a rotary automatic machine. It is assumed that Δt is the time of the displacement of machining process on each workstation of a rotary machine. This Δt parameter is a component of the equation's productivity rate for the rotary machine of parallel structure that should be expressed by the technological and technical data. The following steps develop the equation of the productivity rate for a rotary machine:

The rotor-type automatic machine is processing p work parts or products at the time and is represented by the following equation:

$$t_p = T + \Delta t_1 + \Delta t_2 + \Delta t_3 + \cdots + \Delta t_i + \cdots + \Delta t_{ps-1} \tag{3.4}$$

where t_p is procession time of p work parts; T is the cycle time for machining of one product by one workstation; Δt_i is the displacement time of the workstation i work.

The displacement time Δt on each workstation is constant and equal, i.e. $\Delta t_1 = \Delta t_2 = \Delta t_3 = \cdots \Delta t_i = \cdots \Delta t_{ps-1}$. Then, the total time for machining the p work parts on the rotor machine is represented by the following equation:

$$t_p = T + \Delta t(p_s - 1) \tag{3.5}$$

where p_s is the number of parallel workstations.

Substituting Eq. (3.5) into the equation of the productivity rate per cycle of a rotary machine yields the following equation:

$$Q = \frac{p}{t_p} = \frac{p}{T + \Delta t(p_s - 1)} \tag{3.6}$$

where $p = p_s$ is the number of work parts or products machined. All other parameters are as specified earlier.

The time of displacement of machining Δt on each workstation is defined as the ratio of the passed distance l between two workstations by the one workstation to the circular linear speed V of the rotor machine, (Figure 3.13), i.e. $\Delta t = l/V$. The value of the distance l depends on the geometric parameters of the workstations and rotor's design. The tangential speed V of the rotor's workstations depends on the duration of the machining cycle T, on the value of the length of arc circle's perimeter on which the number of workstations $p_{\alpha+\beta} = p_s - p_\gamma$ are disposed and machining process is implemented, where p_γ is the number of workstations disposed at rotor's sector of angle γ, $(p_s - p_\gamma)$ is the number of workstations disposed at the rotor's sector of angle $(\alpha + \beta)$, (Figure 3.15). Then, the tangential speed is defined by the formula $V = l(p_s - p_\gamma)/T$. Substituting defined parameters and transformation enable defining the time of displacement

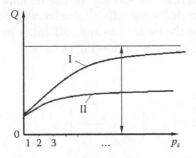

Figure 3.15 The productivity rate of an automated line of parallel structure (I) and rotor-type machine (II) versus the number of workstations.

for machining the work part or product in the workstation that is represented by the following equation:

$$\Delta t = \frac{l}{l(p_s - p_\gamma)/T} = \frac{T}{p_s - p_\gamma} \tag{3.7}$$

where all parameters are specified earlier.

Substituting Eq. (3.7) into Eq. (3.6) and transformation yield the equation of the productivity rate for a rotary automatic machine as follows:

$$Q_T = \frac{p}{T + \dfrac{T}{p_s - p_\gamma}(p_s - 1)} = \frac{p}{T\left(1 + \dfrac{p_s - 1 - p_\gamma + p_\gamma}{p_s - p_\gamma}\right)} = \frac{p}{(t_m + t_a)\left(2 + \dfrac{p_\gamma - 1}{p_s - p_\gamma}\right)} \tag{3.8}$$

where all parameters are specified earlier.

The equation of the total productivity rate for the rotor-type machine of parallel structure is defined with Eq. (3.2) by replacement of expression for the productivity per cycle time and adding to Eq. (3.3) the failure rate of feeder.

$$Q = \frac{p}{(t_m + t_a)\left(2 + \dfrac{p_\gamma - 1}{p_s - p_\gamma}\right)} \times \frac{1}{1 + m_r\left(p_s\lambda_s + \lambda_f + \lambda_{cs}\right)} \tag{3.9}$$

where λ_f is the failure rate of feeder for the rotor machine, $d = \left(2 + (p_\gamma - 1)/(p_s - p_\gamma)\right)$ is the time displacement factor for workstations work. All other parameters are as specified earlier.

Analysis of Eq. (3.9) demonstrates increase in the number of parallel stations p_s leading to an increase in the productivity rate per cycle time and to a decrease in the value of the availability for the rotor machine. It means, there is some limit of productivity rate for the rotor-type machine. The maximum productivity rate is defined by solving the mathematical limit for Eq. (3.9).

The maximal productivity rate for the rotor-type machine of parallel structure depends on the cycle time and reliability indices of one workstation and is represented by the following equation:

$$\lim_{p_s \to \infty} Q = \lim_{p_s \to \infty}\left(\frac{p}{(t_m + t_a)\left(2 + \dfrac{p_\gamma - 1}{p_s - p_\gamma}\right)} \times \frac{1}{1 + m_r\left(p_s\lambda_s + \lambda_f + \lambda_{cs}\right)}\right)$$

$$= \frac{1}{2(t_m + t_a)m_r\lambda_s}$$

$$Q_{max} = \frac{1}{2(t_m + t_a)m_r\lambda_s}$$ (3.10)

where all other parameters are as specified earlier.

Analysis of Eqs. (3.3) and (3.10) of the maximal productivity rate for the automatic machine of linear structure and for the rotor-type machine demonstrates that the latter one gives twice less productivity rate than the automatic machine of linear structure. This result is correct for a greater number of parallel workstations. For small number of workstations, the difference in productivity rate is less, but significant. Figure 3.15 depicted by Eqs. (3.3) and (3.10) demonstrates differences in the productivity rate for the automated line of parallel structure and in the rotor-type machine that is increased with the number of parallel workstations.

Recommendation for the optimal number of parallel workstations in the rotor-type machine is same as for automated line of parallel structure represented earlier. The tendency of change in parameters of the productivity rate for the rotor-type machine is same as for the automated line of parallel structure, but different in values. Increase in the number of parallel workstations in the rotary machine leads to a decrease in the value of the availability and increase in the productivity rate.

A working example

The rotor-type machine of parallel structure and workstations designed with the technical and timing data referred to one product is shown in Table 3.4. The cyclic and total productivity rates and availability for the given data of the rotor-type machine are calculated.

Solution: Substituting the initial data of Table 3.4 into Eqs. (3.10), (3.9) and transformation yield the following results.

The productivity rate per cycle time is as follows:

Table 3.4 Technical data of the rotor-type machine of parallel structure

Title	Data
Machining time, t_{mo} (min)	1.0
Auxiliary time, t_a (min)	0.2
Failure rate of workstation, λ_m	0.0125
Failure rate of feeder, λ_f	4.0×10^{-6}
Failure rate control system, λ_{cs}	6×10^{-8}
Mean repair time, m_r (min)	2.0
Number of workstations, p_s	10
Number of workstations at sector γ, p_γ	2

$$Q_T = \frac{p}{(t_m + t_a)\left(2 + \dfrac{p_\gamma - 1}{p_s - p_\gamma}\right)} = \frac{10}{(1.0 + 0.2)\left(2 + \dfrac{2 - 1}{10 - 2}\right)} = 3.92 \text{ work part/min}$$

The availability is as follows:

$$A = \frac{1}{1 + m_r\left(p_s\lambda_s + \lambda_f + \lambda_{cs}\right)} = \frac{1}{1 + 2.0\left(10 \times 0.0125 + 4.0 \times 10^{-6} + 6.0 \times 10^{-8}\right)}$$

$$= 0.78$$

The total productivity rate of the rotor-type machine is as follows:

$$Q = \frac{p}{(t_m + t_a)\left(2 + \dfrac{p_\gamma - 1}{p_s - p_\gamma}\right)} \times \frac{1}{1 + m_r\left(p_s\lambda_s + \lambda_f + \lambda_{cs}\right)} = 3.92 \times 0.78$$

$$= 3.05 \text{ work part/min}$$

The maximal productivity rate of the rotor-type machine is as follows:

$$Q_{max} = \frac{1}{2(t_m + t_a)m_r\lambda_s} = \frac{1}{2(1.0 + 0.2) \times 3.0 \times 0.0125}$$

$$= 11.1 \text{ work part/min}$$

The results of calculations represented in Table 3.5 enable producing the diagram of changes in the productivity rate the rotor-type machine versus the number of parallel workstations (Figure 3.16).

Table 3.5 Productivity rate for a rotor-type machine with p workstations

p	4	8	16	24	32	40	48	56
Q (work part/min)	1.400	2.614	4.481	5.880	6.972	7.841	8.550	9.150

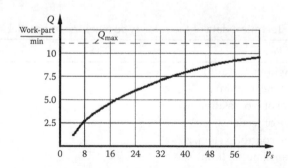

Figure 3.16 Productivity rate for a rotor-type automatic machine with parallel structure versus the number of parallel workstations.

Figure 3.16 shows that the productivity rate increases and asymptotically reaches the limit of Q_{max} = 11.1 work part/min. The optimal number of parallel workstations should be defined by economic approaches for the rotor-type automatic machine.

3.5 Comparative analysis of productivity rates for manufacturing systems of parallel arrangements

Diversity of engineering solutions for the manufacturing systems of parallel structure, which is designed to solve the increase in the productivity rate, can be evaluated by several criteria. Of course, the productivity rate is the primary criterion, with the others being cost of manufacturing system, occupation of production area, design complexity, service system, etc.

These variations of technical, technological and managerial approaches demonstrate that the final decision about the manufacturing system remains in favour of economic efficiency. Nevertheless, industrial practice demonstrates the first approach in the malty stepped evaluation of the manufacturing system and is implemented by the productivity rate index. So far as the weight of productivity rate index among others is prevailed in economic evaluation, the following comparative analysis is used to define the appropriate engineering solution.

Based on the mathematical models for the productivity rate of manufacturing systems of parallel structure, Sections 3.1–3.4 with working examples compute the values of productive outputs. For comparative analysis, the technological, technical data are represented in Table 3.6. Substituting initial data into equations of productivity rates of manufacturing systems of parallel structure defined the values of productivity output (Table 3.7).

Table 3.6 Technological and technical data of manufacturing machines of parallel structure

Title	Data
Machining time, t_{mo} (min)	1.0
Auxiliary time, t_a (min)	0.2
Failure rate of workstation, λ_s	1.25×10^{-2}
Failure rate of feeder, λ_f	4.0×10^{-6}
Failure rate control system, λ_{cs}	6×10^{-8}
Mean repair time, m_r (min)	2.0
Number of workstations, p_s	10
Number of workstations at sector γ, p_γ	2

Table 3.7 Productivity rates of manufacturing machines of parallel structure

No	Manufacturing system	Equation of productivity rate	Productivity rate, Q (work parts/min)
1	Independent workstations	$Q = \dfrac{p}{t_m + t_a} \times \dfrac{1}{1 + m_r(\lambda_s + \lambda_f) + \lambda_{cs}}$	8.130
2	Automated line	$Q = \dfrac{p}{t_m + t_a} \times \dfrac{1}{1 + m_r[p_s(\lambda_s + \lambda_f) + \lambda_{cs}]}$	6.666
3	Rotor-type machine	$Q = \dfrac{p}{(t_m + t_a)\left(2 + \dfrac{p_\gamma - 1}{p_s - p_\gamma}\right)}$ $\times \dfrac{1}{1 + m_r\left(p_s\lambda_s + \lambda_f + \lambda_{cs}\right)}$	3.137

Figure 3.17 illustrates the results of productivity rate in graphical method, where abscissa represents the following: 1) manufacturing system of parallel structure with an independent work of machines; 2) an automated line of parallel structure; 3) a rotor-type machine. The values of productivity rate for manufacturing machines of parallel structure are observed clearly through the histogram with bars.

The productivity rates are arranged accordingly in descending order in Pareto chart, which helps to analyse the data easily and provide a clearer view of results. The comparative analysis of the productivity rates for manufacturing machines of parallel arrangements is implemented, irrelative to the specific area of their applications in industries. Highest productivity rate (Q_1) gives the manufacturing system of parallel structure with

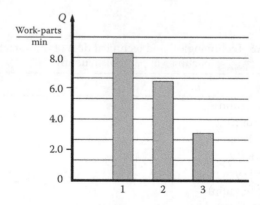

Figure 3.17 Productivity rate in histogram of manufacturing systems of parallel structure.

independent workstations and lowest (Q_3) gives the rotor-type machine. Nevertheless, in most cases, manufacturers prefer to use the rotor-type machine for machining products due to economic reasons in spite of its lowest productivity rates.

Bibliography

1. Badiru, A.B., and Omitaomu, O.A. 2011. *Handbook of Industrial Engineering Equations, Formulas, and Calculations*. Taylor & Francis. New York.
2. Benhabib, B. 2005. *Manufacturing: Design, Production, Automation, and Integration*. 1st ed. Taylor & Francis. New York.
3. Birolini, A. 2007. *Reliability Engineering: Theory and Practice*. 5th ed. Springer. New York.
4. Chryssolouris, G. 2006. *Manufacturing Systems: Theory and Practice*. 2nd ed. Springer. New York.
5. Crowson, R. 2006. *Assembly Processes: Finishing, Packaging, and Automation, Handbook of Manufacturing Engineering*. 2nd ed. Taylor & Francis. New York.
6. Groover, M.P. 2013. *Fundamentals of Modern Manufacturing: Materials, Processes, and Systems*. 5th ed. (Lehigh University). John Wiley. Hoboken, NJ.
7. Kalpakjian, S., and Schmid, S.R. 2013. *Manufacturing Engineering & Technology*. 7th ed. Pearson. Cambridge.
8. Nof, S.Y. 2009. *Springer Handbook of Automation*. Purdue University. Springer-Verlag. Berlin, Heidelberg.
9. Rao, R.V. 2011. *Advanced Modeling and Optimization of Manufacturing Processes*. 1st ed. Springer Series in Advanced Manufacturing. Springer. New York.
10. Shaumian, G.A. 1973. *Complex Automation of Production Processes*. Mashinostroenie. Moscow.
11. Shell, R.L., and Hall, E.L. 2000. *Handbook of Industrial Automation*. Marcel Dekker, Inc. New York.
12. Usubamatov, R., Abdulmuin, Z., Nor, A., and Murad, M.N. 2008. Productivity rate of rotor-type automated lines and optimization of their structure. *Proceedings Institution of Mechanical Engineers, Part B: Journal of Engineering Manufacture*. 222(11). pp. 1561–1566.
13. Usubamatov, R., Ismail, K.A., and Shah, J.M. 2012. Mathematical models for productivity and availability of automated lines. *International Journal of Advanced Manufacturing Technology*. DOI 10.1007/s00170-012-4305-y.
14. Volchkevich, L.I. 2005. *Automation of Production Processes*. Mashinostroenie. Moscow.

chapter four

Manufacturing systems of serial arrangement and productivity rate

Manufacturing processes of the work parts with complex design can be performed on a single machine tool that gives low productivity rate. To increase the productivity rate, the complex machining processes are decomposed uniformly and balanced on several short consecutive operations that are performed on serial workstations. Such production processes are arranged on manufacturing lines of serial arrangements with linear and circular designs. Circular design is preferable for a small number of serial workstations as it is more compact and convenient for maintenance. Automated production lines of a serial structure are most prevalent for the mass manufacturing of work parts, because it occupies less production area and is compact in design than other manufacturing lines of serial arrangement. Such positive properties of the automated lines are the neighbouring ones with negative properties. The failure of one workstation in the automated lines leads to a stop of the entire line with workable other workstations and units and decrease in the productivity rate. To solve this problem, the segmentation of automated lines of serial structure into sections with embedded buffers of different capacities is used, which enable to enhance the productivity rate. The mathematical models for maximal productivity rates of the automated lines with optimal number of serial workstations and buffers of different capacity were formulated.

4.1 Introduction

In engineering, any kind of machines are assembled from different machine parts with different designs like housing, shafts, gears, clutches, bearings, bolts, etc. Most of the machine parts have complex design with high quality of surfaces and accuracy of geometrical parameters. These type of machine parts can be manufactured by sophisticated and long technological processes that can contain tens and more of operations. The total machining time t_{mo} for manufacturing a machine part with complex design is represented as a sum of machining times of separate operations. This total time for machining a work part is expressed by the following

Figure 4.1 Typical machine parts with complex design.

equation: $t_{mo} = \sum_{i=1}^{q} t_i$ where t_i is the machining time of i operation; q is the number of operations. Figure 4.1 represents typical machine parts with complex design like an engine block, multi-stepped shafts with gears and splines, housings for supports, etc.

Manufacturing processes of the work parts with complex design is implemented on the multi-operational machine tools with several sequence of machining processes. However, such processes are performed for a long time on different machine tools and do not give high capacity. To increase the productivity of machining processes of work parts or products, the complex machining processes are decomposed on several short consecutive operations. Each operation is implemented on one machine tool that is designed for execution of simple and short operations. Therefore, a complex technological process is performed on several consecutive machine tools q of an independent work that are arranged as a manufacturing serial line according to the sequence of the technological process. Such method allows to dramatically increase the productivity rate of a manufacturing system. However, this process is linked with other problems like increasing the complexity and production area occupation, cost of production systems, design the means for transportation of work part machined, increasing the numbers of employees that depends on level of automation, etc.

The following analysis of the properties of different types of manufacturing systems of serial arrangements enables for developing mathematical models to calculate the manufacturing system's productivity rate. The productivity rate and reliability of manufacturing system of serial arrangement are changed with the changes in the number of serial workstations. The dependency of the productivity rate from the number of workstations in serial structures of manufacturing systems is more complex than in parallel structures. In industries, manufacturing systems of serial arrangement different in layout and designs are represented by the following solutions:

- Manufacturing lines of serial arrangement with independent work of machine tools or workstations.
- Automated lines of serial structure.
- Automated lines of serial structure segmented on sections with embedded buffers.

These three types of manufacturing systems have different properties and mathematical models for the productivity rate. Detailed considerations of mathematical models for each type of serial manufacturing systems and automated lines are presented below.

4.2 Productivity rate of manufacturing systems with serial arrangement and independent work of workstations

In industrial areas, manufacturing serial lines with independent work of machine tools (Figure 4.2), the transportation of the work parts from one machine tool to other are performed by simple means like rollers, sliders, chute, conveyors, etc. or by storing in small containers (Figure 4.3).

Symbolic picture of the manufacturing line of serial arrangement with independent work of q machine tools is represented in Figure 4.4.

Segmentation of complex manufacturing processes on short operations that are performed by q consecutive machine tools enables increasing the productivity rate by decreasing the value of machining time at machine tools. The average machining time on each machine tool can be expressed by the following equation: $t_{av} = t_{mo}/q$, where t_{mo} is total machining for processing the work part. All short operations are uniformly balanced on the consecutive machine tools that represent the ground for creation of the manufacturing lines of serial structure. The following steps can express all common information that are necessary for designing the manufacturing lines with prescribed properties:

Figure 4.2 Manufacturing systems with independent work of machine tools.

- The complex technological process for machining the work part with complex design is segmented on short operations and distributed uniformly on several serial machine tools according to the sequence of technological process.
- The short operation times on each serial workstation are balanced and enable for defining the necessary number of serial workstations in the manufacturing line.
- Uniform segmentation of the technological process is a complex problem and always there is a bottleneck machine tool with longest machining time for processing the work part.
- The sequence of the machine tools is arranged as a serial manufacturing line with linear structure and independent work of each machine tool.
- The transfer of the machined work part from one machine tool to the other is conducted by simple means of mechanisation (slider, roller, trolley, etc.)
- Failure of any machine tool in the manufacturing line serial arrangement does not lead to stop other machine tools.

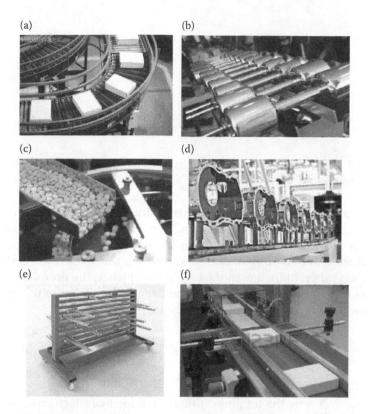

Figure 4.3 Transfer the machined products by typical mechanisms: (a) roller, (b) slider, (c) chute, (d) chain conveyor, (e) trolley and (f) belt conveyor.

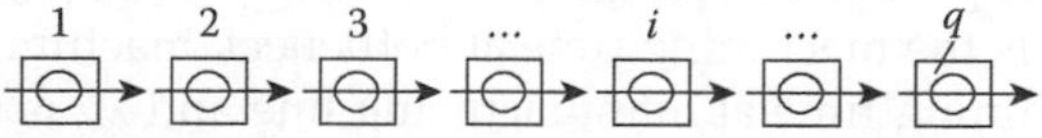

Figure 4.4 Symbolic presentation of the manufacturing line of serial arrangement.

- Productivity rate of the manufacturing line of serial arrangement is determined by the bottleneck machine tool with the longest machining time.
- Manufacturing systems with separate work of machine tools occupy the largest production area.

The following diagram (Figure 4.5) with the average T_{av} and longest machining time represents the work of the manufacturing line of serial arrangement. Latter one belongs to the bottleneck machine tool that determines the cycle time T of the manufacturing line.

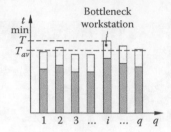

Figure 4.5 Typical diagram of work and balancing for a manufacturing and automated line of serial structure.

The properties of the manufacturing line of serial arrangement represented earlier enable for developing the mathematical model for productivity rate based on Eq. (3.4) for a single machine tool. Equation (3.4) is modified and new components of machining time described earlier are added and inherent for a manufacturing line. Hence, the productivity rate of manufacturing line of serial arrangement with independent a work of machine tools is presented by the following equation:

$$Q = \frac{1}{\dfrac{t_{mo}}{q} f_c + t_a} \times \frac{1}{1 + m_r \lambda_{sb}} \tag{4.1}$$

where $t_{mo} = \sum\limits_{i=1}^{g} t_i$ is the total machining time for manufacturing the work part; t_i is the machining time on i machine tool after balancing the technological process; q is the number of a serial machine tool in the line; t_{mo}/q is the average machining time on each machine tool; $f_c = t_b/(t_{mo}/q)$ is the correction factor, where t_b is the machining time at bottleneck machine tool; $t_{av} = t_{mo}/q$ is average machining time at manufacturing line and λ_{sb} is the failure rate of the bottleneck machine tool. All other parameters are as specified earlier.

Component of Eq. (4.1) that is the availability $A = 1/(1 + m_r \lambda_{sb})$ is considered only for one machine tool, because the productivity rate of manufacturing line of serial arrangement is defined for one bottleneck machine tool. Other serial machines q are implementing the technological processes of work parts with values of the machining time less than at the bottleneck machine tool. Each station is implementing the component of technological process that leads to a decrease of the machining time for the manufacturing system with linear arrangement. This statement is represented in Eq. (4.1) by expression t_{mo}/q that expresses a decrease of the total machining time proportional to the number of serial machine tools.

As mentioned earlier in practical terms, after balancing, a manufacturing line always contains a bottleneck machine tool whose machining time is the longest. However, Eq. (4.1) developed for the average machining time that should be corrected with reference to the longest machining time. Correction factor f_c is added into Eq. (4.1) that expressed the difference in the machining time between the bottleneck machine tool and average machining time of the manufacturing system.

Analysis of Eq. (4.1) enables for depicting the change in productivity rate of the manufacturing line of serial arrangement with increase in the number of serial machine tools in the line. This change is represented in Figure 4.6, which demonstrates increase in the productivity rate for a manufacturing line of serial arrangement versus the number of machine tools in the line. The theoretical maximal productivity rate Q_{max} that can give the manufacturing line of serial structure is defined by solving the mathematical limits of Eq. (4.1).

$$\lim_{q \to t_{min}} Q = \lim_{q \to t_{min}} \left(\frac{1}{\dfrac{t_{mo}}{q} f_c + t_a} \times \frac{1}{1 + m_r \lambda_{sb}} \right) = \frac{1}{(t_{min} + t_a)(1 + m_r \lambda_{sb})} \tag{4.2}$$

where t_{min} is the minimal machining time at the machine tool that decomposition of technological process enables physically for a manufacturing line.

Practically, the maximum productivity rate of a manufacturing line of serial arrangement is limited by the minimised machining time at a bottleneck machine tool, auxiliary time and reliability parameters of the serial machine tool.

A working example

Manufacturing line of serial arrangement with independent a work of machine tool is designed with the technical and timing data referred

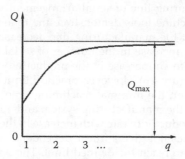

Figure 4.6 The productivity rate of a manufacturing line of serial arrangement versus the number of serial machine tools.

Table 4.1 Technical data of a manufacturing line

Title	Data
Total machining time, t_{mo} (min)	26.0
Auxiliary time, t_a (min)	0.3
Correction factor, f_c	1.2
Failure rate of machine tool, λ_s	0.025
Mean repair time, m_r (min)	3.0
Number of serial machine tools, q	12

to one product shown in Table 4.1. The cyclic and total productivity rates and availability for the given data of the manufacturing line of serial arrangement are calculated.

Solution: Substituting the initial data of Table 4.1 into Eq. (4.1) and the calculating yield the following results.

The productivity rate per cycle is as follows:

$$Q = \frac{1}{\dfrac{t_{mo}}{q} f_c + t_a} = \frac{1}{\dfrac{26}{12} \times 1.2 + 0.3} = 0.4 \text{ work part/min}$$

The availability is as follows:

$$A = \frac{1}{1 + m_r \lambda_s} = \frac{1}{1 + 3.0 \times 0.025} = 0.93$$

The total productivity rate of the manufacturing line of serial arrangement is as follows:

$$Q = \frac{1}{\dfrac{t_{mo}}{q} f_s + t_a} \times \frac{1}{1 + m_r \lambda_s} = 0.4 \times 0.93 = 0.37 \text{ work part/min}$$

The results of calculation of the productivity rate and parameters for the manufacturing line of serial arrangement with an independent work of machine tools demonstrate the following. Increase in the failure rate of the manufacturing line decreases the magnitude of its availability. Increase in the number of serial machine tools in the system leads to an increase in the productivity rate for a manufacturing line (Figure 4.6). However, practical limit of the productivity rate depends on the minimal machining time, reliability of the machine tools in the manufacturing system and auxiliary time. The increase of the productivity rate with increase in the number of serial machine tools does not give a linear dependency and has a monotonical tendency to reach the defined limit. The optimal number of serial machine tools in the manufacturing line of linear arrangement with the independent work of workstations should be defined by the economical criterion.

4.3 Productivity rate of an automated line with serial structure

Automated production lines of a serial structure are most prevalent in the mass manufacturing of the products and work parts. Designers of automated lines joined serial workstations with hard linking and single transport and control systems. Such construction of the automated line occupies less production area than manufacturing lines with independent serial machine tools. Contemporary practice demonstrates that automated lines of serial structure is designed by two schemes: first is the linear design and second is circular design. Both of them have own properties. Circular design is preferable for small number of serial workstations, as it is more compact and convenient for maintenance.

Automated lines of circular arrangement are equipped by rotary table, where the located work parts are machined. This rotary table implements function of the transport mechanism for all work parts. The rotary table performs the indexed angular motion and displaces all work parts to other sequence workstations that locate at the periphery of the rotary table.

Automated lines of linear structure are equipped by transport systems with linear motion. This transport system displaces all work parts to be machined on one step to workstations and repeats transporting them after each cycle time of machining process.

In the automated lines, the failure of one workstation leads to stop the entire automated line with workable other workstations, mechanisms and units. This attribute in case of increase in the number of serial workstations leads to a decrease in the productivity rate of the automated line of serial structure. Such problem brings forward the task of structural optimisation of the automated line by the criterion of maximal productivity rate, which yields the optimal number of serial workstations. Figure 4.7 presents the view of typical automated lines of serial structure that can be found in different industries.

A symbolic picture for the automated line of serial structure with q workstations is represented in Figure 4.8.

Segmentation and balancing of the technological process on the sequence q serial workstations is conducted as for manufacturing line of serial structure discussed earlier. The properties inherent to the automated line of serial structures are represented later:

- The sequence of the workstations is arranged as serial automated line with hard linking and equipped by single transport and control systems.
- Special designed linear or circular transport mechanism and units conduct the transfer of the work parts from one workstation to the other.

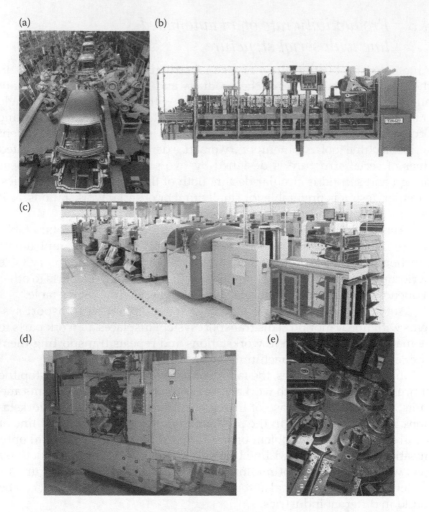

Figure 4.7 Automated lines of serial structure: (a) car welding line with robots, (b) packaging line, (c) electronic components processing line, (d) multi-spindle machine tool (in short, an automated line of circular arrangement), (e) multi-workstation automated line of circular arrangement.

- Failure of any workstation, mechanisms or units in the automated line of serial structure leads to a stop of other workstations and the entire line.
- Productivity rate of the automated line of serial structure is determined by the bottleneck workstation with longest machining time and reliability indices of all workstations, mechanisms and units.
- Automated line of serial structure occupies less production area than other types of manufacturing systems.

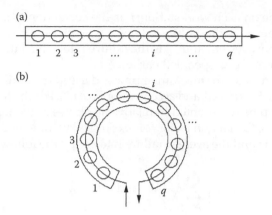

Figure 4.8 Symbolic presentation of the automated line of serial structure with q workstations and linear (a) and circular (b) arrangements.

The properties of the automated line of serial structure presented earlier enable developing the mathematical model of productivity rate based on Eqs. (4.2) and (2.8) for a single machine tool or workstation. Combining the known properties and equations for the productivity rate of manufacturing system and adding new properties of automated line of serial structure is developed by the equation of productivity rate for the automated line with q serial workstations.

The industrial practice demonstrates that the reliability attributes of serial workstations of automated lines are different. Variety of failure rate that causes decreasing productivity rate and bottleneck machining time demonstrate specifically in mathematic form and represent the sustainable solution for productivity improvement for the automated line of serial structure. Then, the productivity rate of automated line with serial workstations and linear structure is represented by the following equation [13]:

$$Q = \frac{1}{\dfrac{t_{mo}}{q} f_c + t_a} \times \frac{1}{1 + m_r(\lambda_{s1} + \lambda_{s2} + \lambda_{s3} + \cdots + \lambda_{si} + \cdots + \lambda_{sq} + \lambda_{tr} + \lambda_{cs})}$$

$$= \frac{1}{\dfrac{t_{mo}}{q} f_c + t_a} \times \frac{1}{1 + m_r\left(\displaystyle\sum_{i=1}^{q} \lambda_{si} + \lambda_{tr} + \lambda_{cs}\right)} \qquad (4.3)$$

where t_{mo} is the total machining time for manufacturing the work part; q is the number of a serial workstations in the line; t_{mo}/q is the average machining time on each workstation; f_c is the correction factor for longest machining time; λ_{si} is the failure rate of a workstation i; λ_{tr} is the failure rate of a transport system and λ_{cs} is the failure rate of a control system. All other parameters are as specified earlier.

A correction factor for machining time is derived in detail for the manufacturing line in Section 4.1 and accepted for Eq. (4.3). For the following analysis, Eq. (4.3) can be represented in relative attributes of the failure rates for the workstations. Correction factor f_{si} is expressed by ratio of the failure rate of a workstation i and the average failure rate $\lambda_{s.av}$ by the following equation:

$$\lambda_{sav} = \left(\sum_{i=1}^{q} \lambda_{si} \right)/q \quad \text{and} \quad f_{si} = \lambda_{si}/\lambda_{sav}$$

Substituting defined expressions into Eq. (4.3) and transformation yield the equation for productivity rate of the automated line of serial structure:

$$Q = \frac{1}{\dfrac{t_{mo}}{q}f_c + t_a} \times \frac{1}{1 + m_r[\lambda_{sav}(f_{s1} + f_{s2} + f_{s3} + \cdots + f_{si} + \cdots + f_{sq}) + \lambda_{tr} + \lambda_{c.s}]}$$

$$= \frac{1}{\dfrac{t_{mo}}{q}f_c + t_a} \times \frac{1}{1 + m_r(\lambda_{sav}\sum_{i=1}^{q} f_{si} + \lambda_{tr} + \lambda_{cs})}$$

(4.4)

where all components are as specified earlier.

In an ideal situation, when failure rate of all workstations are equal, the correction factors for failure rate equal to 1.0, i.e. $f_{si} = \lambda_{si}/\lambda_{sav} = 1.0$. In this case, the productivity rate for the automated line of serial structure is presented by the following equation [9,14]:

$$Q = \frac{1}{\dfrac{t_{mo}}{q}f_c + t_a} \times \frac{1}{1 + m_r(q\lambda_{sav} + \lambda_{tr} + \lambda_{c.s})}$$

(4.5)

where all parameters are specified earlier.

Practically, the serial automated line with equal failure rates for all workstations is a very rare event. Hence, Eq. (4.5) can be used for rough estimation and calculations at the preliminary steps of design an automated line of serial structure. Analysis of Eqs. (4.3–4.5) enables depicting the change in the productivity rate of the automated line of serial structure with increase in the number of serial workstations. Equations (4.3–4.5) have two components in denominator, for which the values are changing in opposite directions with change in the number of serial workstations q and other components are not changeable. It means that Eqs. (4.3–4.5) have extreme functions that can be defined graphically or analytically by the rules applied for differential equations. These changes in variable components of Eq. (4.5) are represented in Figure 4.9 and demonstrate increase in the productivity rate for an automated line of serial structure versus the number of serial workstations.

Analytically, the optimum number of serial workstations for the automated line of serial structure by the criterion of maximum productivity rate is obtained by solving the first derivative of Eq. (4.5) with respect to the numbers of serial workstations and setting it to zero. The first derivative results in the following equation:

$$\frac{dQ}{dq} = \frac{d\left(\dfrac{1}{\dfrac{t_{mo}}{q}f_c + t_a} \times \dfrac{1}{1 + m_r(q\lambda_{sav} + \lambda_{tr} + \lambda_{cs})} \right)}{dq} = 0$$

giving rise to the following:

$$t_{mo}f_c[1 + m_r(q\lambda_{sav} + \lambda_{tr} + \lambda_{c.s})] - m_r\lambda_{sav}(t_{mo}f_c q + q^2 t_a) = 0$$

or

$$q_{opt} = \sqrt{\frac{t_{mo}f_c[(1/m_r) + \lambda_{cs} + \lambda_{tr}]}{t_a\lambda_{sav}}} \tag{4.6}$$

Defined equation of the optimal number of serial workstations is not necessary and sufficient condition for the judgement of the productivity rate Q value. To judge the Q value for maximum productivity, the second

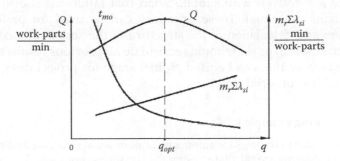

Figure 4.9 Change of the productivity rate parameters versus change of number of workstations in the automated line of serial structure.

derivative of Eq. (4.6) is needed to find the optimal number of serial work-stations q. The second derivative of Eq. (4.6) is presented by the following equation:

$$\frac{d^2 Q}{dq^2} = \frac{d\left(t_{mo} f_c [1 + m_r (\lambda_{tr} + \lambda_{c.s})] - m_r \lambda_{sav} t_a q^2\right)}{dq^2} = 0$$

giving rise to the following:

$-2 m_r \lambda_{sav} t_a q = 0$, i.e. the negative result.

The second derivative giving the negative result, i.e. $d^2 Q/dq^2 < 0$ then the graph of Eq. (4.6) is concave up, it means the function of productivity rate Q has maximal value. The optimal number of workstations in the automated line of serial structure giving maximal productivity rate that presented by Eq. (4.6) depends on the failure rates of workstation, transport and control systems, mean repair time and bottleneck machining and auxiliary time. Substituting Eq. (4.6) into Eq. (4.5) and transformation yield the equation of the maximal productivity rate for the automated line of serial structure:

$$Q_{max} = \frac{1}{\sqrt{\dfrac{t_{mo} f_c t_a \lambda_{sav}}{(1/m_r) + \lambda_{cs} + \lambda_{tr}}} + t_a}$$
$$\times \frac{1}{1 + m_r \left\{ \sqrt{\dfrac{\lambda_{sav} t_{mo} f_c [(1/m_r) + \lambda_{cs} + \lambda_{tr}]}{t_a}} + \lambda_{tr} + \lambda_{cs} \right\}} \qquad (4.7)$$

As mentioned earlier, Eqs. (4.6) and (4.7) of the optimal number of work-stations and the maximal productivity rate for the automated line of serial structure are derived with simplification that failure rates of the serial workstations are equal. These equations can be used for preliminary estimation and calculation of the structure of the serial automated line. However, Eqs. (4.6) or (4.7) should recalculate all previous results and give accurate answer that can be used at final stage for project design of the automated line of serial structure.

A working example 1

The productivity rates for automated lines of serial structure and different designs are calculated using the equations represented earlier. The following sections describe the analysis of the productivity rate for typical designs of the automated line of serial structure. For simplicity of calculation, it is accepted that the failure rates of groups of workstations are equal. An engineering problem is to calculate the productivity rates and demonstrate the change in productivity rate in terms of change in the number of serial workstations.

The automated line of serial structure can be designed with several variants in terms of $q = 20, ..., 30$ serial workstations (Figure 4.8a). The basic technical and technological data for the automated line of serial structure are represented in Tables 4.2 and 4.3.

Solution: Calculations are conducted by two variants. Equations (4.5) and (4.7) conduct preliminary calculations that enable for estimating the potential maximal productivity rate and the optimal number of workstations [9,14]. Final calculations are conducted using Eq. (4.4), which enables for getting accurate results that can be used for practical design of the automated line of serial structure.

Preliminary calculations. Substituting initial data of Tables 4.2 and 4.3 into Eqs. (4.5) and (4.7), the calculated results are represented by the following:

The optimal number of serial workstations:

$$q = \sqrt{\frac{t_{mo} f_c [(1/m_r) + \lambda_c + \lambda_{tr}]}{t_a \lambda_{sav}}}$$

$$= \sqrt{\frac{35.0 \times 1.2 \times [(1/3.0) + 8.0 \times 10^{-4} + 4 \times 10^{-5}]}{0.3 \times 7.0 \times 10^{-2}}} \approx 26$$

Table 4.2 Technical data for the automated line of serial structure

Title	Data
Total machining time, t_{mo} (min)	35
Auxiliary time, t_a (min)	0.3
Number of stations, q	20, ..., 30
Correction factor for the bottleneck station f_c	1.2

Table 4.3 Reliability indices for the automated line

Title	q	$\lambda_{s,i}$ (failures/min)
Failure rate of the workstations q, λ_s	1–5	7.0×10^{-2}
	6–10	5.0×10^{-2}
	11–15	8.0×10^{-2}
	16–20	6.0×10^{-2}
	21–25	9.0×10^{-2}
	26–30	7.0×10^{-2}
	Average	7.0×10^{-2}
Failure rate of the control system, λ_c		8.0×10^{-4}
Failure rate of the transport system, λ_{tr}		4.0×10^{-5}
Mean repair time, $m_r = 3.0$ min		

The maximal productivity rate

$$Q = \frac{1}{\dfrac{t_{mo}}{q}f_c + t_a} \times \frac{1}{1 + m_r(q\lambda_{sav} + \lambda_{tr} + \lambda_{c.s})}$$

$$= \frac{1}{\dfrac{35.0}{26} \times 1.2 + 0.3} \times \frac{1}{1 + 3.0(26 \times 0.07 + 8.0 \times 10^{-4} + 4.0 \times 10^{-5})}$$

$$= 0.08 \text{ work part/min}$$

The productivity rate per cycle time is as follows:

$$Q_c = \frac{1}{\dfrac{t_{mo}}{q}f_c + t_a} = \frac{1}{\dfrac{35.0}{26} \times 1.2 + 0.3} = 0.522 \text{ work part/min}$$

The availability is as follows:

$$A = \frac{1}{1 + m_r(q\lambda_{sav} + \lambda_{tr} + \lambda_{cs})} = \frac{1}{1 + 3.0(26 \times 0.07 + 8.0 \times 10^{-4} + 4.0 \times 10^{-5})}$$

$$= 0.154$$

Accurate calculations. Substituting initial data of Table 4.2 into Eq. (4.4) with variables of the numbers of workstations q and their failure rates λ_{si}, the following diagram (Figure 4.10) represents incremental calculating results:

$$Q = \frac{1}{\dfrac{t_{mo}}{q}f_c + t_a} \times \frac{1}{1 + m_r\left(\displaystyle\sum_{i=1}^{q}\lambda_{si} + \lambda_c + \lambda_{tr}\right)}$$

$$= \frac{1}{\dfrac{35}{q} \times 1.2 + 0.3} \times \frac{1}{1 + 3.0\left(\displaystyle\sum_{i=1}^{q}\lambda_{si} + 8.0 \times 10^{-4} + 4.0 \times 10^{-5}\right)}$$

The calculated results represented in Table 4.4 are used to depict the diagram of changes in the productivity rate versus the number of serial workstations with different failure rates (Figure 4.10).

Figure 4.10 demonstrates the change in the productivity rate for the automated line of serial structure with a different number of workstations q and failure rates. The diagram depicts the curves calculated accurately by Eq. (4.4) and by Eq. (4.7) with average data. The maximal productivity rate $Q = 0.085$ prod/min describes the automated line

Table 4.4 Productivity rate for a serial automated line with q stations

q	5	10	15	20	25	30
Q (product/min)	0.056	0.079	0.080	0.085	0.080	0.078
Q_{av} (product/min)	0.056	0.071	0.077	0.08	0.082	0.08

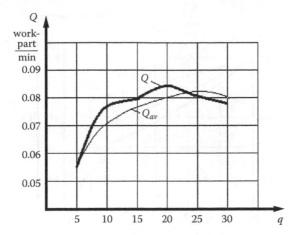

Figure 4.10 Productivity rate of an automated line of serial structure versus the number of workstations with different failure rates.

with $q = 20$ serial workstations (curve Q). The curve Q_{av} represents the change in the productivity rate calculated with the average failure rates of the workstations [9,14]. For the optimal number of serial workstations ($q = 26$), Eq. (4.4) determines the maximal productivity rate of the automated line of serial structure ($Q_{max} = 0.08$ prod/min). The difference in results calculated by Eqs. (4.4) and (4.7) is considerable. Hence, it is obligatory to calculate the productivity rate of industrial machines and systems based on real technical data. A simplified approach in the calculations yields a bigger optimal number of serial workstations that cannot be accepted for final solutions.

A working example 2

The automated line of serial structures (Figure 4.7) can be designed with variants, for which the technical data is presented in Table 4.5.

Table 4.5 Technical data for the automated line of serial structures

Title	Data
Total machining time, t_{mo} (min)	1.0
Auxiliary time, t_a (min)	0.1
Number of workstations, q	10, ..., 30
Correction factor for the bottleneck workstation, f_c	1.2
Reliability attributes for an automated line	
Average failure rate of the workstation, λ_s (fail/min)	$\lambda \times 10^{-3}$
Failure rate of the control system, λ_{cs}	$\lambda \times 10^{-8}$
Failure rate of the transport system, λ_{tr}	$\lambda \times 10^{-10}$
Mean repair time, m_r	3.0 min

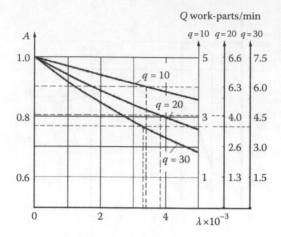

Figure 4.11 Productivity and availability rate of the automated line with q serial workstations versus the increase in the failure rate λ of the automated line.

Substituting the given data into Eq. (4.4), the results for the availability of the automated line of serial structure and its productivity versus the failure rate are presented in Figure 4.11. Figure 4.11 demonstrates the change in availability and productivity rate for the automated line of serial structure with a different number of workstations q versus change in failure rates.

If the accepted productivity rate is $Q = 4.0$ part/min, then:

- An automated line with $q = 10$ serial workstations should have a failure rate of $\lambda \leq 3.4 \times 10^{-3}$ failures/min that gives the availability $A = 0.9$.
- An automated line with $q = 20$ serial workstations should have a failure rate of $\lambda \leq 3.8 \times 10^{-3}$ failures/min that gives the availability $A = 0.805$.
- An automated line with $q = 30$ serial workstations should have a failure rate of $\lambda \leq 3.3 \times 10^{-3}$ failures/min that gives the availability $A = 0.77$.

The structural variants of the automated line give the productivity rate $Q_t = 4.0$ parts/min with a different number of serial workstations and failure rates.

The diagram depicts that increasing the failure rate of the automated line decreases the magnitude of its availability. Increasing the number of serial workstations in the automated line of serial structure leads to decrease in the value of availability. The productivity rate for the line is increased and reached a maximal value for the defined optimal number of workstations and then decreased with the increase in the number of workstations. This variable dependency in productivity rate versus the number of workstations gives extreme solutions.

4.3.1   Industrial case study

The newly developed mathematical model is needed to be validate and analyse the real industry world. The validation has taken place at a motorcycle industry under an assembly line [11]. A company that produces the motorcycle with an assembly line of serial structure in linear arrangement was chosen for this study. This company manifested big interest to proposed research and for the analysis the used mathematical model and statistical method application to validation purpose are required. The mathematical model is suitable to validate a company because the assemble line is semi-automated, where there is a large conveyor for transport system to carry the body of motorcycle from one assembly station to the following one. Although the mathematical model is best applied in fully automated line, the assembly line of mass production is acceptable. The motorcycle, which is complete after final assembly, is shown in Figure 4.12.

The assembly line of motorcycles contains a long linear-type transport system. The chassis of the motorcycle is hanging and carried by the transport system. The assembly process takes place with the support of engineers and technicians of the production system.

A company produces several models of motorcycle, and for research study, one model of motorcycle is selected for detailed analysis and accurate data collection. Retrospective statistical data from previous years regarding reliability, failure rates, idle times and reasons were collected. Observation of the assembly line work within 31 production shifts was conducted and 160 failures or breakdowns that occurred in the assembly line were recorded [11]. The obtained data set satisfied with at least 90% of statistical and probability trustworthiness. Another two sets of data, which are technical, and reliability within 31 shifts are presented in Tables 4.6 and 4.7.

After data collection stage, the collected data are required to proceed to the calculation stage using Eq. (4.4) of the mathematical model for

Figure 4.12 Assembly line of motorcycles.

Table 4.6 Technical data of an assembly line

Title	Data
Total assembly time, tmo (min)	26.55
Auxiliary time, ta (min)	0.52
Number of assembly stations, q	37
Assembly time at the bottleneck station, $t_{m.bt}$ (min)	0.8969
Correction factor for assembly time of the bottleneck station, f_c	1.25
Number of motorcycles assembled, z	7,103
Observation time is 31 shifts, θ (min)	14,760

productivity rate with different failure rates of assembly stations. Equation $Q_{ac} = z/\theta$ is applied to calculate the actual productivity rate for assembly line. Substituting collected data from Tables 4.6 and 4.7 into defined equation yield the following results:

Actual productivity rate : $Q_{ac} = z/\theta = 7013/14760 = 0.475$ motorcycle/min = 28.506 motorcycles/h

Productivity rate with different values of failure rate for an assembly station:

$$Q = \frac{1}{\dfrac{t_{mo}}{q} f_c + t_a} \times \frac{1}{1 + m_r \left(\sum_{i=1}^{q} \lambda_{si} + \lambda_c + \lambda_{tr} \right)} = \frac{1}{\dfrac{26.55}{37} \times 1.25 + 0.52}$$

$$\times \frac{1}{1 + 11.24(7.249 \times 10^{-3} + 2.033 \times 10^{-4} + 9.485 \times 10^{-4} + 3.048 \times 10^{-2})}$$

$$= 0.491 \text{ motorcycle/min}$$

Percentage error for theoretical Q with actual Q_a productivity rate is as follows:

$$\delta = \frac{Q - Q_a}{Q} = \frac{0.491 - 0.475}{0.491} \times 100\% = 3.25\%$$

The percentage error, when compared to the result computed by mathematical model and to actual productivity rate result, demonstrates perfect matching. The reason for the difference in result can be explained by not accurately computing the failure rates for the assembly stations and other components. So, it is proved and validated that mathematical model of productivity rate with different station failure rate is accurate. Hence, practitioners in industrial area for practical application can use this model of productivity rate.

The mathematical model for productivity rate of an automated line of serial structure and with different failure rate of workstations enables to compute the productivity losses by different reasons. Peculiarity of this

Table 4.7 Reliability data of the assembly line

Title	Assembly station q number	$\lambda_{s,i}$ (1/min)
Failure rate of assembly stations, q	1	1.355×10^{-4}
	2	8.130×10^{-4}
	3	4.743×10^{-4}
	4	8.130×10^{-4}
	5	0.000
	6	5.420×10^{-4}
	7; 8	0.000
	9	6.775×10^{-5}
	10	4.065×10^{-4}
	11	6.775×10^{-5}
	12 … 16	0.000
	17	4.065×10^{-4}
	18	6.775×10^{-5}
	29; 20	0.000
	21	6.775×10^{-5}
	22 … 24	0.000
	25	7.453×10^{-4}
	26	4.065×10^{-4}
	27	5.420×10^{-4}
	28	5.420×10^{-4}
	29	6.775×10^{-5}
	30	0.000
	31	8.130×10^{-4}
	32	0.000
	33	1.355×10^{-4}
	34….35	0.000
	36	6.775×10^{-5}
	37	6.775×10^{-5}
Total assembly station failure rate, λ		7.249×10^{-3}
Failure rate of the control system, λc		9.485×10^{-4}
Failure rate of the transport system, λtr		2.033×10^{-4}
Failure rate of the defective parts, λd		3.048×10^{-2}
Mean repair time, mr (min)		11.24

mathematical model is the ability for computing the maximal productivity rate and defining optimal structure, i.e. number of serial workstations in the automated line. Application of such analytical model for an assembly line with manual processes is intended as assembly activities of employees are implemented at permanent time with minor variations. This assumption means that the mathematical model for productivity rate derived for automated lines can be used for assembly lines with permanent manual processes. However, there is restriction in defining the potential productivity rate for assembly lines with manual operations. This restriction depends on the physical ability of alive assemblers, in which the speed of motions and force activity cannot be competitive to automatic machines. In such cases, the value of the time for assembly process and auxiliary motions in mathematical models for productivity rate will be bigger and permanent than for machine motions. This limitation of human activity should be accepted for computing the maximal productivity rate of the manufacturing systems with manual processes.

An analysis of failure rates of the assembly line is implemented to define the productivity rate losses that can be studied and listed down for the sake of sustainable improvement of the assembly process. Equations given later represent the calculation of the values of productivity rate losses and Figure 4.13 demonstrates the total productivity loss diagram for the assembly line of motorcycles.

1. Potential productivity rate that depends on physical ability of assemblers. This index is correct if facilities of the assembly line are reliable.

$$Q_c = \frac{1}{\dfrac{t_{mo}}{q}f_c + t_a} = \frac{1}{\dfrac{26.55}{37} \times 1.25 + 0.52} = 0.705 \text{ motorcycle/min}$$

$$= 42.3 \text{ motorcycle/h}$$

2. The productivity rate of the assembly line proportionally decreases with increase in failure rates of assembly stations.

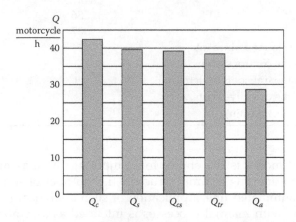

Figure 4.13 The change in productivity rates due to failure rates of the assembly line of motorcycles.

The total productivity loss of the assembly line is as follows:

$$\Delta Q = Q_c - Q_a = 0.705 - 0.475 = 0.230 \text{ motorcycle/min}$$

$$= 13.8 \text{ motorcycle/h}.$$

The total failure rate represents the sum of all failure rates of stations and other units as follows: $\lambda = 3.888 \times 10^{-2}$. Other productivity losses of the assembly line are calculated by the ratios of failure rates represented in Table 4.7 as follows:

$$\lambda_s : \lambda_c : \lambda_{tr} : \lambda_d = (7.249 \times 10^{-3}) : (2.033 \times 10^{-4}) :$$
$$(9.485 \times 10^{-4}) : (3.048 \times 10^{-2})$$

3. The productivity losses due to failure rates of assembly stations:

$$\Delta Q_s = \Delta Q \lambda_s / \lambda = 0.230 \times 7.249 \times 10^{-3} / 3.888 \times 10^{-2}$$
$$= 0.0428 \text{motorcycle/min} = 2.568 \text{ motorcycles/h}$$

Then, the productivity rate of the assembly line is decreased and is represented as follows:

$$Q_s = Q_c - \Delta Q_s = 42.3 - 2.568 = 39.732 \text{ motorcycles/h}$$

4. The productivity losses due to failure rates of control system:

$$\Delta Q_{cs} = \Delta Q \lambda_c / \lambda = 0.230 \times 2.033 \times 10^{-4} / 3.888 \times 10^{-2}$$
$$= 0.0012 \text{ motorcycle/min} = 0.072 \text{ motorcycle/h}$$

Then productivity rate of the assembly line is decreased and represented as follows:

$$Q_{cs} = Q_s - \Delta Q_{cs} = 39.732 - 0.072 = 39.660 \text{ motor cycle/h}$$

5. The productivity losses due to failure rates of transport system:

$$\Delta Q_{tr} = \Delta Q \lambda_{tr} / \lambda = 0.230 \times 9.485 \times 10^{-4} / 3.888 \times 10^{-2}$$
$$= 0.0056 \text{ motorcycle/min} = 0.336 \text{ motorcycle/h}$$

Then productivity rate of the assembly line is decreased and represented as follows:

$$Q_{tr} = Q_{cs} - \Delta Q_{tr} = 39.660 - 0.336 = 39.324 \text{ motorcycle/h}$$

6. The productivity losses due to failure rates of defected assembly of motorcycles:

$$\Delta Q_d = \Delta Q \lambda_d / \lambda = 0.230 \times 3.048 \times 10^{-2} / 3.888 \times 10^{-2}$$
$$= 0.1803 \text{ motorcycle/min} = 10.818 \text{ motorcycles/h}$$

Then, the actual productivity rate of the assembly line is decreased and represented as follows:

$$Q_a = Q_{tr} - \Delta Q_d = 39.324 - 10.818 = 28.506 \text{ motorcycle/h}$$

The decrease in the productivity rate due to failure rates of mechanisms and units of the assembly line can be observed clearly in Pareto chart (Figure 4.13). Pareto chart helps to analyse the data easily and provide a clearer view of results and priority directions for sustainable improvement of the motorcycle assembly line.

The diagram in Figure 4.13 and equations represent all productivity rate losses as result of mathematic calculation. Highest productivity rate losses due to defected assembly of motorcycles, which is $\Delta Q_d = 10.818$ motorcycle/h in assembly line. Second sensitive productivity losses are caused due to assembly stations failure rate in the line, which contribute $\Delta Q_s = 2.568$ motorcycle/h. These two types of productivity losses represent challenges for engineers to find technical solutions that can decrease such manufacturing flaw. Other types of productivity losses are caused due to failure rates of control systems and transport with $\Delta Q_{cs} = 0.072$ and $\Delta Q_{tr} = 0.336$ motorcycle/h respectively, i.e. their reliability is more or less suited to production process.

After completely analysing the overall failure rate that causes loss of productivity rate, assembly stations within the assembly line are sometimes required to determine their optimal number that gives the maximal productivity rate. This problem is solved by simple numerical computing of Eq. (4.4) with the variable of the number of assembly stations that enables the representation of the graphical diagram. This diagram can show the maximal productivity rate of the assembly line that gives the optimal number of assembly stations. For the given assembly line, this problem is not to be placed, because this line is designed for the necessary productivity rate.

4.4   Productivity rate of an automated line segmented on sections with buffers of limited capacity

The technological processes of mass production manufacturing systems are implemented on multi-station automated lines. Industrial experience shows that automated lines do not always operate at their maximal capacity due to a number of reasons. The automated lines experience downtimes for different reasons while actual productivity rates are often lower than those that calculated by the known mathematical models and operating manuals. Various methods have been used in the past to improve the productivity rate of automated lines. One of them is the segmentation of automated lines into sections with embedded buffers, which is supposed to enhance the productivity rate.

Numerous publications in the area of technological processes have shown that operating of automated lines still has many unresolved problems. Manufacturers need mathematical models for the productivity rate of production systems with complex designs, such as the automated lines based on the technology of processes, the reliability of mechanisms and machines of production systems, the number of sections with buffers, etc. These mathematical models should predict the actual productivity rate of industrial machines and systems, while the difference between theoretical and practical output should be minimal. This is crucial task for practitioners of manufacturing engineering. This chapter contains an analytical model for determining the physical productivity rate of an automated line that is segmented into sections with embedded buffers and with workstations and sections of different failure rates [13]. This model should solve the problems related to an increase in productivity rate by calculating the optimal number of workstations for any given number of buffered sections that provides maximal productivity rate of automated lines.

The balancing of technological processes and implementing on multistations arrangements in automated lines is not endless and restricted by many reasons like quality of machining the work parts, reliability of workstations, etc. In addition, practice and previous study demonstrate that the productivity rate of automated lines increased with increase in the number of workstations until defined limit. Following the increase one leads to decrease in the productivity rate of automated lines. There is an optimal number of workstations that gives maximum productivity. Reliability of workstations and other units restricts to add more stations and leads to increase the downtime of automated lines. Increase in the productivity of automated lines is possible by the segmentation of the line into sections with embedded buffers, which are supposed to enhance the productivity rate. In this case, segmentation of automated lines on several sections leads to reducing the downtimes of entire line. The downtime of any workstation will not only stop the entire automated line but also stop only one section while other sections continue to work and filling and taking work parts from intersectional buffers. This engineering solution can increase the productivity rate of the automated lines and represents interests for industries.

Practically, the buffer capacity of the automated lines can be limited by different reasons like restriction of production space, design problems, cost of facilities, etc. In such cases, the intersectional buffers can compensate partially the downtimes of the neighbouring sections due to failure rates. Proposed mathematical model is solving the problem related to an increase in productivity rate by calculating the optimal number of workstations for any given number of buffered sections that provide maximum productivity rate of automated lines.

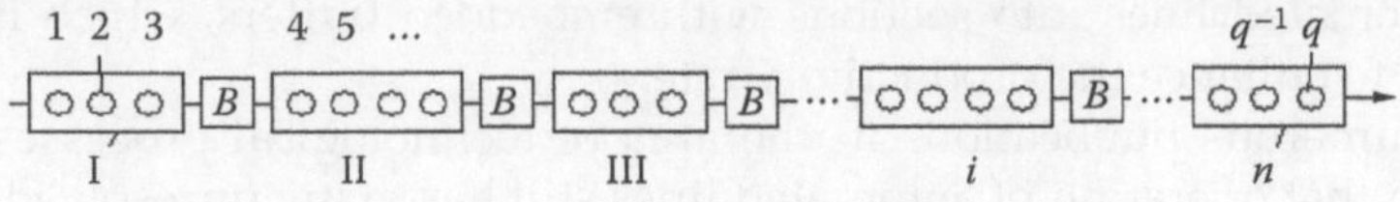

Figure 4.14 Scheme of an automated line with q stations, segmented into n sections with B buffers.

Known mathematical models of the manufacturing system productivity rate are based on the average data of machining time, the reliability attributes of workstations and the machine units of an automated line. These analytical models can give approximate results that should be corrected by actual technical data based on the technological process, the design of an automated line and the real industrial environment. The equation for the productivity rate of automated lines with hard-linked workstations and mechanisms is derived and represented earlier. This equation demonstrates that failure of one workstation leads to downtime in the other one due to hard linkage. Segmentation of an automated line and placing buffers between sections allow for reducing the downtime of the line, leading to an increase in the productivity rate. Schematically, this automated line of serial structure segmented on sections is represented in Figure 4.14.

Modern manufacturing machines and systems with complex designs require trustful mathematical models to compute their productivity rate. The balancing of the technological process leads to an increase in the number of stations q and to a decrease in the cycle time T. The productivity rate equation of an automated line segmented into sections with buffers is represented by the following modified equation:

$$Q = \frac{1}{\dfrac{t_{mo}}{q} f_c + t_a} \times \frac{1}{1 + m_r \left(f_s \dfrac{\sum\limits_{i=1}^{q} \lambda_{s.i}}{n} + \sum\limits_{i=1}^{n-1} \Delta\lambda_i + \lambda_{bf} + \lambda_c + \lambda_{tr} \right)} \qquad (4.8)$$

where t_{mo} is the total machining time for processing the work part; q is the number of workstations; $t_{av} = t_{mo}/q$ is the average machining time on the workstation; $f_c = t_{mb}/t_{av}$ is the correction factor of the machining time t_{mb} for the bottleneck workstation; m_r is the mean repair and service time of failure of the mechanism and units; $\lambda_{s.i}$ is the failure rate of workstation i; n is the number of sections in an automated line,

$\lambda_m = \left[\left(\sum\limits_{i=1}^{q} \lambda_{s.i} \right) / n \right] + \lambda_{bf} = \left[\left(\sum\limits_{i=1}^{n} \lambda_i \right) / n \right] + \lambda_{bf}$ is the mean failure rate of one section, λ_j is the failure rate of section i; $\sum_{i=1}^{n-1} \Delta\lambda_i$ is the added failure rates of sections due to limited capacity of buffers; the automated line designed with $(n-1)$ buffers; λ_b is the failure rate of the buffer; λ_c and λ_{tr} are the failure rates of the control system and transport mechanism, respectively.

Practically, it is impossible to have an equal failure rate for each section. There is always a bottleneck section with a high level of failure rate, which determines the productivity rate of an automated line. The mean failure rate λ_m of one section as represented earlier is corrected by the failure rate λ_b of the bottleneck section. Hence, the failure rate λ_b of the bottleneck section is represented by the expression

$$\lambda_b = \left[\left(\sum_{i=1}^{q} \lambda_{s.i}\right)\Big/n\right]f_s + f_b = \left[\left(\sum_{i=1}^{n} \lambda_i\right)\Big/n\right]f_s + f_b,\ \text{where}\ f_s = \lambda_b/\lambda_m\ \text{is the cor-}$$

rection factor that expresses the difference between the failure rate of the bottleneck section and the average failure rate of the section.

In a real production environment, the buffer with limited capacity increases the downtimes of the neighbouring sections. Each section has its own downtime and added uncompensated downtime of the failed section due to the limited capacity of the buffers that are located between sections. Then, as the neighbouring sections fail, they increase the downtime or failure rate of other sections and decrease the productivity rate of an automated line. The nature of failure rates in an automated line and stations of sections are different and random. The number of workstations that failed at the same time is variable and random. The probability of coincidence of failed workstations is exponentially decreasing with increase in the number of workstations.

The capacity of the buffer is determined by the possible downtime of the section that is based on the probabilistic approach for the failures of workstations. The coincided failures of workstations lead to increase in the section's downtime that should be compensated by the capacity of the buffer and do not transfer to the neighbouring sections. The total number of failure rates of the section represents the sum of the failure rates of the workstations and units. However, the fails of several workstations can coincide with the time of maintenance by the technician of the one failed workstation. This condition is expressed by decreasing the total number of failures of the section. Practically, the probability of coincided fails of several workstations for a short time has a very small value of high order that can be neglected.

Analysis and computing of the productivity rate of an automated line with sections and buffers of limited capacity are based on the real data and the following approaches:

- Failure rates of workstations of an automated line are different.
- An automated line is segmented into sections for which the failure rates are different.
- An automated line always has the bottleneck section with the high failure rate than others.
- Buffers have identical design, reliability indices and limited capacity.
- Buffer's limited capacity leads to additional failure rate in the neighbouring sections, which are represented by the added failure rate.

The current workability of an automated line with a group of workstations is supported by the team of technicians that conduct the maintenance of mechanisms, units, etc. and repair of its random failures. The number of workstations of an automated line in maintenance by one technician is represented by the normative data that depend on the reliability of workstations and other units that are distinguished for different industries. The mean time to repair of random failures in metal-cutting industries is represented by the short time that enables removing the failure of workstation and other mechanisms. Except this, the technician has 30% of the time for inspection, control and regular support of the workability of workstations.

Capacity of the buffer is solved on the basis of technician's maintenance time for an automated line, its cycle time and minimum downtime of workstations. The technicians maintain the workstations in which fails coincide. This condition is important for computing the uncompensated failure rates. The total number of failures of the automated line should be evenly distributed among the number of technicians. However, when the technician maintains one failed workstation, but at the same time, other workstations can be failed and be out of his service, and this leads to increasing the downtime of sections and the entire automated line of serial structure. The duration of the downtime of the section determines the capacity of the buffer that fills up or feeds the work parts of the section. If the cycle time of an automated line is T, then the necessary capacity of the buffer that compensates the downtime of the section and do not add to the additional downtime of other sections is expressed by the following equation:

$$N = \frac{\theta}{T} = \frac{q_s \times m_r}{T} \tag{4.9}$$

where N is the capacity of the buffer or the total number of the work parts located in the buffer for filling or feeding the section; θ is the downtime of the section; T is the cycle time; q_s is the number of all workstations in service by one technician; m_r is the mean time to repair failures and service of the workstation and other components are as specified earlier.

Practically, the buffer can be of limited capacity due to design reason. Then, the failure rates of the sections are compensated partially. This condition leads to the additional downtimes of the neighbouring sections, i.e. the uncompensated failure rates are transferred and added to failure rates of other sections. These additional failure rates are defined by the following equation:

$$\Delta \lambda_{i-1} = \lambda_i \left(1 - \frac{N_l}{N} \right) \tag{4.10}$$

where $\Delta \lambda_{i-1}$ is the additional failure rates of the neighbouring section; λ_i is the failure rates of the section of the automated line and N_l is the limited capacity of the buffer. All other components are as specified earlier.

The buffer's capacity is defined with considering of the probabilistic fails and works of the workstations and sections. The workstation's downtimes of the section coincide randomly. The probability of all workstations of independent work will have downtime at the same time and is calculated as the product of probabilities of failures of all workstations. The downtime for probabilistic consideration of the failures is accepted for maintenance of workstations that is in service by one technician. In engineering, the probability of density distribution of failures is represented commonly by the exponential function. The probability that the workstation i will work or have failure during a defined time of work is represented by the following equations:

$$R_i(t) = e^{-\lambda_i t}$$
$$P_i(t) = 1 - R_i(t) = 1 - e^{-\lambda_i t} \tag{4.11}$$

where $R_i(t)$ is the probability that workstation, section or some unit i will work during a defined time (t) without failures; $P_i(t)$ is the probability that workstation, section or some unit i will have failure during a defined time (t) of work; λ_i is the failure rate of the workstation, section or some unit i; t is the work time of the automated line of serial structure, and all other parameters are as specified earlier.

The probability of coincided failures at the same time $P_q(t)$ of q workstations with different failure rates is presented by the following equation:

$$P_q(t) = (1 - e^{-\lambda_1 t}) \times (1 - e^{-\lambda_2 t}) \times (1 - e^{-\lambda_3 t}) \times \cdots \times (1 - e^{-\lambda_q t}) \tag{4.12}$$

The limited volume of probability of the fail at one time for several workstations that leads to sensitive downtime of the section is considered for computing the capacity of the buffer. The limited capacity of the buffer enables filling or pick up of the work parts by the sections while the failed section is fixed, but not in full. The buffer of limited capacity compensates the downtime of the failed section partially. This situation leads to forced downtimes of the workable neighbouring sections that has the right dependency on the value of the failure rates of sections. The value of added failures $\Delta\lambda_i$ represents an increase in the downtimes of sections and shows which part of the uncompensated failure rates of sections is added to the neighbouring sections of an automated line.

The increase in the section's downtime is defined analytically by the following scheme. All sections of an automated line have additional failure rates due to overfilling or lack of work parts in the buffers of limited capacity. Analysis of increase in the section's downtime is considered for a common case, the bottleneck section i is located somewhere in the middle of the line. The failure rate of the sections located in front and behind of the failed section i is increased by adding the uncompensated failure rates of the bottleneck section. Hence, the total failure rates of the sections are calculated by two approaches.

First approach is the added failure rates of the sections located in front (f_i) of the bottleneck section one due to overfilling the buffers by work parts (Figure. 4.8, work parts enter to the automated line). For simplicity of analysis, the failure rates of the bottleneck section i is expressed as λ_i. The added failure rates, due to limited capacity of the one buffer, to the sections located in front of the bottleneck section are represented as follows:

- $\Delta\lambda_{f(i-1)}$ is the added failure rates to the first section from the bottleneck section i.
- $\Delta\lambda_{f(i-2)} + \Delta\lambda_{f(i-1)}$ is the added failure rates to second section from the two foregoing sections, where $\Delta\lambda_{f(i-2)}$ is the added failure rates from the first section to the second section.

Hence, the added failure rates to the last section k of the automated line from the sections located in front of the bottleneck one due to the limited capacity of buffers will have the following expression:

$$\Delta\lambda_{fk} = \Delta\lambda_{f(i-k)} + \cdots + \Delta\lambda_{f(i-2)} + \Delta\lambda_{f(i-1)} \tag{4.13}$$

where all parameters are as specified earlier.

Second approach considers the added failure rates of the sections located behind (b_i) the bottleneck section due to lack of work parts in the buffers. The added failure rates to the sections that are located behind the bottleneck section is defined by a similar method as represented earlier. Hence, the added failure rates to last section n from the sections located behind the bottleneck one due to the limited capacity of buffers will have the following expression:

$$\Delta\lambda_{bn} = \Delta\lambda_{b(i+1)} + \Delta\lambda_{b(i+2)} + \cdots + \Delta\lambda_{b(i+n)} \tag{4.14}$$

The total added failure rates to the sections located in front and behind of the bottleneck one are obtained by combining Eqs. (4.13) and (4.14) and transformation. Therefore, $\Delta\lambda = \Delta\lambda_{fk} + \Delta\lambda_{bn}$ or

$$\sum_{i=1}^{n-1} \Delta\lambda_i = \Delta\lambda_{f(i-k)} + \cdots + \Delta\lambda_{f(i-2)} + \Delta\lambda_{f(i-1)} + \Delta\lambda_{b(i+1)} + \Delta\lambda_{b(i+2)} + \cdots + \Delta\lambda_{b(i+n)}$$

(4.15)

where each component of added failure rates of the sections is expressed by Eq. (4.10).

Substituting Eq. (4.15) into Eq. (4.8) yields the following equation:

$$Q = \frac{1}{\frac{t_{mo}}{q} f_c + t_a} \times \frac{1}{\left[1 + m_r \dfrac{f_s \sum\limits_{i=1}^{q} \lambda_{s.i}}{n} + \Delta\lambda_{f(i-k)} + \cdots + \Delta\lambda_{f(i-2)} + \Delta\lambda_{f(i-1)} + \Delta\lambda_{b(i+1)} + \Delta\lambda_{b(i+2)} + \cdots + \Delta\lambda_{b(i+n)} + \lambda_{bf} + \lambda_c + \lambda_{tr} \right]}$$

(4.16)

where i is the bottleneck section and other parameters are as specified earlier.

Equation (4.16) can be solved manually or using a software for accurate estimation of the productivity rate for considered structure of an automated line. The productivity rate for an automated line with buffers after each workstation ($n = q$) is also represented by Eq. (5.17). Analysis of Eq. (4.16) demonstrates that for the given number of sections the serial automated line can be designed with optimal number of workstations, which gives the maximal productivity rates. The solution for the optimisation of the automated line structure is the same as represented earlier in Section 4.2.

A working example

An automated line segmented on sections with buffers of limited capacity can be designed with variants. An engineering problem is stated to define the changes in productivity rate with the change in the number of sections. The segmentation of an automated serial line into sections leads to a decrease in failure rates and an increase in productivity rate. This statement is demonstrated for an automated line that can be segmented on several sections with buffers. The productivity rate is considered for segmentation of an automated line from two until five sections. The basic technical data of an automated line are represented in Tables 4.8 and 4.9. For simplicity of calculations and to avoid cumbersome mathematical expressions, it is accepted that the failure rates of workstations are equal. The automated line contains

Table 4.8 Technical data of an automated line segmented on
sections with buffers

Title	Data
Machining time, t_{mo}, (min)	35
Correction factor for the bottleneck workstation, f_c	1.2
Auxiliary time, t_a, (min)	0.2
Number of workstations, q	30
Cycle time, $T = [(t_{mo}q)/f_c] + t_a$, (min)	0.95
Number of sections, n	2, ..., 5
Capacity of the buffer, N	8
Number of workstations in service by one technician, q_s	6
Number of technicians servicing an automated line	5

Table 4.9 Reliability indices of an automated line

Title	Section	Number of workstations	$\lambda_{s.i}$ of workstation	λ_i of section
			Mean number of failures/min	
Failure rate of Two sections	1	14	0.0403	0.5642
	2	16	0.0396	0.6336
Three sections	1	10	0.0388	0.3880
	2	10	0.0420	0.4200
	3	10	0.0333	0.3330
Four sections	1	7	0.0458	0.3206
	2	7	0.0400	0.2800
	3	9	0.0375	0.3375
	4	7	0.0378	0.2646
Five sections	1	5	0.0440	0.2200
	2	6	0.0416	0.2496
	3	7	0.0382	0.2674
	4	6	0.0375	0.2250
	5	6	0.0390	0.2340
Failure rate of the control system, λ_c				8.0×10^{-5}
Failure rate of the transport system, λ_{tr}				4.0×10^{-5}
Failure rate of the buffer, λ_{bf}				3.0×10^{-7}
Total failure rate λ				2.23592
Mean failure rate of the workstation, λ_s				0.03967
Mean repair and service time, m_r (min)				2.6

30 workstations in service by a team of technicians. The maximum downtime of the section is when simultaneously failed six workstations need to be serviced by one technician.

1. The productivity rate of an automated line with two sections and one buffer. The equation of the productivity rate for the line with two sections and one buffer is represented by the modified Eq. (4.16).

$$Q = \frac{1}{\dfrac{t_{mo}}{q} f_c + t_a} \times \frac{1}{1 + m_r \left[\dfrac{f_s \sum\limits_{i=1}^{n} \lambda_i}{n} + \Delta \lambda_{f(i-1)} + \lambda_{bf} + \lambda_c + \lambda_{tr} \right]} \tag{1}$$

where all components are as specified earlier.

For solution at this line, the second bottleneck section is accepted with the failure rate $\lambda_2 = 0.6336$ (Table 4.9). Based on the data of Tables 4.8 and 4.9 and equations, the following parameters for Eqs. (4.8) and (4.10) are computed: The correction factor for a section is as follows:

$$f_s = \frac{\lambda_{bt}}{\left[\left(\sum\limits_{i=1}^{n} \lambda_i \right) \Big/ n \right]} = \frac{0.6336}{(0.5642 + 0.6336)/2} = 1.0579$$

The probability of failure of six workstations at one time that is serviced by one technician is computed by Eq. (4.12). The downtime of the section for probabilistic consideration of the failures is as follows: $t = m_r q_s = 2.6 \times 6 = 15.6$ min. The probability of the fail at one time of six workstations at the bottleneck section is as follows (Eq. 4.12):

$$P_q(t) = (1 - e^{-\lambda_1 t})^q = (1 - e^{-0.0396 \times 15.6})^6 = 0.00957$$

This probability is very low and should not be taken into consideration. In the manufacturing engineering is accepted the rule, if the result of some processes gives more than 5%–10% of the change, it should be taken into account for consideration followed by an analysis and a solution. The probability of the fail at one time of four workstations is as follows (Eq. 4.12):

$$P_q(t) = (1 - e^{-\lambda_1 t})^q = (1 - e^{-0.0396 \times 15.6})^4 = 0.0450$$

The probability of the fail at one time of three workstations is more than 0.101. Hence, the full capacity of the buffer that compensates downtime of bottleneck section is computed for the four workstations as follows:

$$N = \frac{q_s \times m_r}{T} = \frac{4 \times 2.6}{0.95} = 11$$

The limited capacity of the buffer is accepted as $N_l = 8$ due to design reason.

Substituting defined parameters into Eq. (4.10) and computing yield the added failure rates to the first section of an automated line in front of the bottleneck section:

$$\Delta\lambda_{f(i-1)} = \lambda_i\left(1 - \frac{N_l}{N}\right) = 0.6336 \times \left(1 - \frac{8}{11}\right) = 0.1728$$

Substituting the defined data into Eq. (1) and transforming yield the following result:

$$Q = \frac{1}{\frac{35}{30} \times 1.2 + 0.2} \times \frac{1}{1 + 2.6 \times \left[\begin{array}{c} \dfrac{1.0579 \times (0.5642 + 0.6336)}{2} + 0.3240 + \\ 3.0 \times 10^{-7} + 8.0 \times 10^{-5} + 4.0 \times 10^{-5} \end{array}\right]}$$

$$= 0.179 \text{ work parts/min.}$$

2. The productivity rate of an automated line with three sections and two buffers.

The equation of the productivity rate for the line with three sections and two buffers is represented by the modified Eqs. (4.10) and (4.16).

$$Q = \frac{1}{\frac{t_{mo}}{q}f_c + t_a} \times \frac{1}{1 + m_r\left[\dfrac{f_s\sum_{i=1}^{n}\lambda_i}{n} + \Delta\lambda_{f(i-1)} + \Delta\lambda_{b(i+1)} + \lambda_{bf} + \lambda_c + \lambda_{tr}\right]} \quad (2)$$

where all components are as specified earlier.

This line has the second bottleneck section with the failure rate $\lambda_2 = 0.4200$ (Table 4.9). Then, the correction factor of a section is as follows:

$$f_s = \lambda_{bt}\Big/\left[\left(\sum_{i=1}^{n}\lambda_i\right)\Big/n\right] = \frac{0.4200}{(0.3880 + 0.4200 + 0.3330)/3} = 1.1042$$

Substituting defined parameters into Eq. (4.10) and computing yield the added failure rates to the first sections of an automated line in front and behind of the bottleneck section:

$$\Delta\lambda_{f(i-1)} = \Delta\lambda_{b(i+1)} = \lambda_i\left(1 - \frac{N_l}{N}\right) = 0.4200 \times \left(1 - \frac{8}{11}\right) = 0.1145$$

Substituting the defined data into Eq. (4.16) and transforming yield the following result:

$$Q = \frac{1}{\frac{35}{30} \times 1.2 + 0.2} \times \frac{1}{\left[\frac{1.1224 \times (0.3206 + 0.2800 + 0.3375 + 0.2646)}{4} + 1 + 2.6 \times \left| 0.1014 + 0.0920 + 0.0920 + 3.0 \times 10^{-7} + 8.0 \times 10^{-5} + 4.0 \times 10^{-5} \right| \right]}$$

= 0.239 work parts/min

3. The productivity rate of an automated line with four sections and three buffers.

The equation of the productivity rate for the line with three sections and two buffers is represented by the modified Eqs. (4.12) and (4.16).

$$Q = \frac{1}{\frac{t_{mo}}{q} f_c + t_a} \times \frac{1}{\left[1 + m_r \left[\frac{f_s \sum_{i=1}^{n} \lambda_i}{n} + \Delta\lambda_{f(i-2)} + \Delta\lambda_{f(i-1)} + \Delta\lambda_{b(i+1)} + \lambda_{bf} + \lambda_c + \lambda_{tr} \right] \right]} \tag{3}$$

where all components are as specified earlier.

This line has the third bottleneck section with the failure rate $\lambda_3 = 0.3375$ (Table 4.9). Then, the correction factor for a section is as follows:

$$f_s = \lambda_{bt} \left/ \left[\left(\sum_{i=1}^{n} \lambda_i \right) \middle/ n \right] \right. = \frac{0.3375}{(0.3206 + 0.2800 + 0.3375 + 0.2646)/4} = 1.1224$$

Substituting defined parameters into Eq. (4.10) and computing yield the added failure rates to the first sections of an automated line in front and behind of the bottleneck section:

$$\Delta\lambda_{f(i-1)} = \Delta\lambda_{b(i-1)} = \lambda_i \left(1 - \frac{N_l}{N} \right) = 0.3375 \times \left(1 - \frac{8}{11} \right) = 0.0920$$

The total failure rates of the first section in front of the bottleneck section is as follows:

$$\lambda_{f.t.(i-1)} = \lambda_{f.(i-1)} + \Delta\lambda_{f(i-1)} = 0.2800 + 0.0920 = 0.3720$$

This value is used for calculation of the added failure rates for the following sections. The added failure rates to the second section of the automated line of serial structure in front of the bottleneck section is as follows:

$$\Delta\lambda_{f(i-2)} = \lambda_{f.i(i-1)}\left(1 - \frac{N_l}{N}\right) = 0.3720\times\left(1 - \frac{8}{11}\right) = 0.1014$$

Substituting the defined data mentioned earlier into Eq. (4.16) and transformation yield the following result:

$$Q = \frac{1}{\dfrac{35}{30}\times 1.2 + 0.2}\times\frac{1}{\left[1 + 2.6\times\begin{array}{l}\dfrac{1.1224\times(0.3206 + 0.2800 + 0.3375 + 0.2646)}{4} + \\[2mm] 0.1014 + 0.0920 + 0.0920 + \\[2mm] 3.0\times 10^{-7} + 8.0\times 10^{-5} + 4.0\times 10^{-5}\end{array}\right]}$$

= 0.239 work parts/min

4. The productivity rate of an automated line with five sections and four buffers.

   The equation of the productivity rate for the line with five sections and four buffers is represented by the modified Eqs. (4.12) and (4.16).

$$Q = \frac{1}{\dfrac{t_{mo}}{q}f_c + t_a}\times\frac{1}{1 + m_r\left[\dfrac{f_s\sum\limits_{i=1}^{n}\lambda_i}{n} + \Delta\lambda_{f(i-2)} + \Delta\lambda_{f(i-1)} + \Delta\lambda_{b(i+1)} + \Delta\lambda_{b(i+2)} + \lambda_{bf} + \lambda_c + \lambda_{tr}\right]}$$

(4)

where all components are as specified earlier.

This line has the third bottleneck section with the failure rate $\lambda_3 = 0.2674$ (Table 4.9). Then, the correction factor for a section is as follows:

$$f_s = \lambda_{bt}\bigg/\left[\left(\sum_{i=1}^{n}\lambda_i\right)\bigg/n\right] = \frac{0.2674}{(0.2200 + 0.2496 + 0.2674 + 0.2250 + 0.2340)/5} = 1.1178$$

Substituting defined parameters into Eq. (4.10) and computing yield the added failure rates to the first sections of an automated line in front and behind of the bottleneck section:

$$\Delta\lambda_{f(i-1)} = \Delta\lambda_{b(i-1)} = \lambda_i\left(1 - \frac{N_l}{N}\right) = 0.2674\times\left(1 - \frac{8}{11}\right) = 0.0729$$

The total failure rates of the first sections in front and behind of the bottleneck section, respectively, are as follows:

$$\lambda_{f.t.(i-1)} = \lambda_{f.(i-1)} + \Delta\lambda_{f(i-1)} = 0.2496 + 0.0729 = 0.3225$$
$$\lambda_{b.t.(i+1)} = \lambda_{b.(i+1)} + \Delta\lambda_{b(i+1)} = 0.2250 + 0.0729 = 0.2979$$

This value is used for calculation of the added failure rates for the following sections. The added failure rates to the second section of the automated line of serial structure in front and behind of the bottleneck section, respectively, are as follows:

$$\Delta \lambda_{f(i-2)} = \lambda_{f.i(i-1)} \left(1 - \frac{N_l}{N}\right) = 0.3225 \times \left(1 - \frac{8}{11}\right) = 0.0879$$

$$\Delta \lambda_{b(i+2)} = \lambda_{b.i(i+1)} \left(1 - \frac{N_l}{N}\right) = 0.2979 \times \left(1 - \frac{8}{11}\right) = 0.0812$$

Substituting defined data into Eq. (4.16) and transforming yield the following result:

$$Q = \frac{1}{\dfrac{35}{30} \times 1.2 + 0.2} \times \frac{1}{1 + 2.6 \times \left[\dfrac{\dfrac{1.1545 \times (0.03967 \times 30)}{30} + 0.3200 +}{3.0 \times 10^{-7} + 8.0 \times 10^{-5} + 4.0 \times 10^{-5}}\right]}$$

$$= 0.320 \text{ work parts/min}$$

Following analysis of the added values of the failure rates demonstrates that, after five sections, the total added failure rates to the sections does not increase and can be accepted as constant. For the given example, the total added failure rates for the six and more sections are accepted as $\Delta \lambda = 0.3200$. Then, the change in productivity rate with change in the number of the following sections is accepted as almost of linear dependency. The final structure is an automated line with buffers after each workstation, for which productivity rate is computed by Eq. (4.16). It is accepted that the automated line has the bottleneck workstation with the failure rate $\lambda_{s.bt} = 0.0458$ and mean failure rate $\lambda_s = 0.03967$ (Table 4.9). This data is used for computing the correction factor for a section as follows:

$$f_s = \lambda_{bt} \Big/ \left[\left(\sum_{i=1}^{q} \lambda_s\right) \Big/ n\right] = \frac{0.0458}{(0.03967 \times 30)/30} = 1.1545$$

Substituting defined data into Eq. (4.16) and transforming yield the following result:

$$Q = \frac{1}{\dfrac{35}{30} \times 1.2 + 0.2} \times \frac{1}{1 + 2.6 \times \left[\dfrac{\dfrac{1.1224 \times (0.3206 + 0.2800 + 0.3375 + 0.2646)}{4} +}{\begin{matrix} 0.1014 + 0.0920 + 0.0920 + \\ 3.0 \times 10^{-7} + 8.0 \times 10^{-5} + 4.0 \times 10^{-5} \end{matrix}}\right]}$$

$$= 0.239 \text{ work parts/min}$$

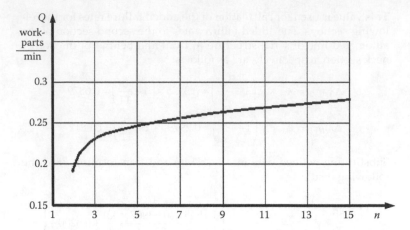

Figure 4.15 Productivity rate of an automated line versus the number of sections with buffers of limited capacity.

The obtained result in the productivity rate of an automated line with the changes in the number of sections was computed by Eq. (4.16) and was represented in Figure 4.15.

> The diagram of the changes in the productivity rates (Figure 4.15) demonstrates that the added failure rates to the sections due to limited capacity of buffers reflect intensively on the output of the automated line with the 4–5 numbers of sections. The following computing demonstrates that increase in the number of sections in an automated line leads to a proportional increase in the productivity rate.

4.4.1 An automated line segmented on sections with buffers of big capacity

Practically, manufacturing processes for small sizes of work parts implemented on automated lines segmented on sections are designed with buffers of big capacities. The small sizes of work parts enable designing the buffers for which the capacities can fully compensate downtimes of neighbouring sections. This constructional peculiarity of the buffers and their properties can be represented in a modified mathematical model for the productivity rate of automated lines with buffers derived in Section 4.3. The buffers that compensate downtimes of neighbouring sections do not create additional failure rates for other sections of the automated line. In such case, Eq. 4.16 is simplified, where components of added failure rates to the sections and the correction factor for a section are omitted from equation. Then, the mathematical model for productivity rate of automated line segmented on sections with buffers of big capacities is represented by the following equation:

$$Q = \frac{1}{\frac{t_{mo}}{q} f_c + t_a} \times \frac{1}{1 + m_r \left(f_s \dfrac{\sum\limits_{i=1}^{q} \lambda_{s.i}}{n} + \lambda_{bf} + \lambda_c + \lambda_{tr} \right)}$$

(4.17)

$$= \frac{1}{\frac{t_{mo}}{q} f_c + t_a} \times \frac{1}{1 + m_r (\lambda_{sb} + \lambda_{bf} + \lambda_c + \lambda_{tr})}$$

where $\lambda_{sb} = f_s \left(\sum\limits_{i=1}^{q} \lambda_{s.i}/n \right)$ is the failure rate of the bottleneck section, and other parameters are as specified earlier.

Equation (4.17) is relatively simple and is solved manually for the considered structure of an automated line, including a line with buffers after each workstation ($n = q$). Analysis of Eq. (4.17) demonstrates that, for the given number of sections n, the automated line of serial structure can be designed with optimal number of workstations, which gives the maximal productivity rate.

A working example 1

An automated line segmented on sections with buffers of big capacity can be designed with variants. The productivity rate is considered for an automated line segmented on sections with buffers embedded after each workstation. An engineering problem is stated to define the changes in the productivity rate with the change in the number of automated line sections. The technical data of an automated line are represented in Tables 4.8 and 4.9 (Section 4.3), and all approaches accepted as an example are considered. The following calculations are conducted for Eq. (4.17) to define the productivity rates of the automated line with the change in the number of sections. All intermediate calculations of the parameters for Eq. (4.16) are accepted and used for the variable structure of the automated line.

1. The productivity rate of an automated line with two sections and one buffer.

   Substituting the defined data (Tables 4.8 and 4.9) into Eq. (4.17) and transformation yield the following result:

$$Q = \frac{1}{\frac{35}{30} \times 1.2 + 0.2} \times \frac{1}{1 + 2.6(0.6336 + 3.0 \times 10^{-7} + 8.0 \times 10^{-5} + 4.0 \times 10^{-5})}$$

$$= 0.236 \text{ work parts/min}$$

2. The productivity rate of an automated line with three sections and two buffers. Substituting the defined data into Eq. (4.17) and transformation yield the following result:

$$Q = \frac{1}{\frac{35}{30} \times 1.2 + 0.2} \times \frac{1}{1 + 2.6(0.4200 + 3.0 \times 10^{-7} + 8.0 \times 10^{-5} + 4.0 \times 10^{-5})}$$

= 0.298 work parts/min

3. The productivity rate of an automated line with four sections and three buffers.

   Substituting defined parameters into Eq. (4.17) and computing yield the following result:

$$Q = \frac{1}{\frac{35}{30} \times 1.2 + 0.2} \times \frac{1}{1 + 2.6(0.3375 + 3.0 \times 10^{-7} + 8.0 \times 10^{-5} + 4.0 \times 10^{-5})}$$

= 0.332 work parts/min

4. The productivity rate of an automated line with five sections and four buffers. Substituting defined data into Eq. (4.17) and transformation yield the following result:

$$Q = \frac{1}{\frac{35}{30} \times 1.2 + 0.2} \times$$

$$\frac{1}{1 + 2.6(0.2674 + 3.0 \times 10^{-7} + 8.0 \times 10^{-5} + 4.0 \times 10^{-5})}$$

= 0.368 work parts/min

The following analysis demonstrates that increase in the number of sections yields an almost linear proportional increase in the productivity rate of the automated line segmented on sections and can be accepted as constant. The final structure is an automated line with buffers after each workstation, for which the productivity rate is computed by Eq. (4.17). Substituting defined data into Eq. (4.17) and transforming yield the following result:

$$Q = \frac{1}{\frac{35}{30} \times 1.2 + 0.2} \times \frac{1}{1 + 2.6 \times \left[\frac{0.03967 \times 30}{30} + 3.0 \times 10^{-7} + 8.0 \times 10^{-5} + 4.0 \times 10^{-5} \right]}$$

= 0.622 work parts/min

Obtained result in the productivity rate of an automated line with the changes in the number of sections was computed by Eq. (4.17) and represented in Figure 4.16.

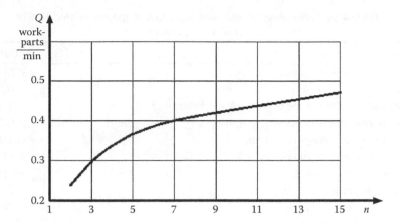

Figure 4.16 Productivity rate of an automated line versus the number of sections with buffers of big capacity.

The diagram of the changes in the productivity rates (Figure 4.16) demonstrates that increase in the productivity rates with segmentation on the sections is intensive for the first steps including 4–5 sections. The following computing demonstrates that increasing the number of sections in an automated line leads to an almost linear proportional increase in the productivity rate.

A working example 2

The buffered automated line of serial structure with the given number of sections n yields the maximal productivity rate Q_{max} and optimal number of serial workstations q_{opt}. Increasing the number of the sections with buffers leads to increasing the number of optimal serial workstations. This statement is proofed by substituting the initial data from Table 4.10 for the automated line segmented on sections with buffers of big capacity into Eq. (4.17) for the given number of sections.

Computed results represented in Table 4.11 and diagrams of productivity rate versus change in the number of serial stations for the given number of sections are represented in Figure 4.17.

Figure 4.17 demonstrates the optimal number of the serial workstations increased with increase in the number of sections of the automated line with serial structure. However, after 5–6 sections of the automated line, the number of optimal workstations increased less intensively, confirmed by the dotted line and Figure 4.16. A similar picture of the change in parameters for the automated line of serial structure was segmented into sections with embedded buffers of limited capacity.

The presented mathematical approach and method for calculating parameters based on the technical and technological data of an

Table 4.10 Technological and technical data of the automated line of serial arrangement

Title	Data
Machining time, t_{mo} (min)	20.0
Auxiliary time, t_a (min)	0.2
Correction factor for the bottleneck workstation, f_c	1.2
Failure rate of workstation, λ_s per min	0.2
Failure rate of transport system, λ_{tr}	4.0×10^{-7}
Failure rate of control system, $\lambda_{c.s}$	6×10^{-8}
Failure rate of buffer, λ_b	6×10^{-8}
Mean repair time, m_r (min)	3.0
Number of workstations, q_s	36
Number of sections, n	3 … 5

Table 4.11 The change in the productivity rate versus change in the number of serial stations and sections

n	q	15	20	25	30	32	34	35	36
2	Q	0.101	0.102	0.101					
3			0.142	0.143	0.1428				
4				0.181	0.182	0.181			
5					0.2173	0.2174	0.2171		
6						0.2504	0.2506	0.2508	0.2506

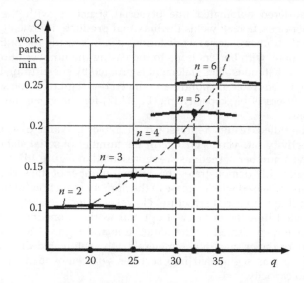

Figure 4.17 The change in the productivity rate versus change in the number of serial stations and sections.

automated line segmented on sections with buffers of limited or big capacity allow defining the number of sections of the line by the criterion of necessary productivity rate. This structure of the production line is typical in manufacturing area, for which the productivity rate is computed by different methods. The obtained mathematical model represents the holistic equation that allows the computing of productivity rate of an automated line with buffers of limited or big capacity and different indices of reliability for section and other units. This mathematical model includes components of technology, auxiliary time, mean repair time and availability. Equations (4.16), (4.17) and (4.4) enable defining the optimal number of workstations for an automated line with the defined number of sections, which ultimately give the maximum productivity rate. The mathematical model also enables the evaluation of structural variations, i.e. the number of workstations and sections based on their failure rates, and the capacity of buffers that gives different productivity rates for an automated line. The equations obtained show that increasing the number of sections yields growth in the productivity rate of an automated line with sections and buffers.

The proposed mathematical model will be useful in modelling the output of an automated line with different reliability indices of sections, workstations and mechanisms. The equation obtained is simple and based on known data in industrial area that enables easy computation of basic parameters of an automated line with a complex structure.

4.5 Comparative analysis of productivity rates for manufacturing systems of serial arrangements

The manufacturing systems of serial arrangements are created for the complex technological processes for machining work parts of sophisticated geometry. These systems are the most prevalent in manufacturing industries. The serial structure of machining systems designed with wide diversity of engineering constructions enables increase in the productivity rate. Great variety in design of manufacturing systems of serial structure implies the same variety of their productivity rates. These variations in designs should be evaluated by the comparative analysis with criterion of productivity rate. Such analysis is the first approach to define the appropriate engineering solution for a manufacturing system.

The mathematical models for the productivity rate of manufacturing systems of serial structure are represented in Sections 4.1–4.3. The working examples and methods for computing the values of productivity rates are used for the following analysis. The comparative analysis of productivity rates is conducted for the defined serial structures of manufacturing systems with one technological process and one serial workstation.

For simplicity of computing, the productivity rate for the manufacturing systems of serial structures and automated line segmented into sections accepted the following properties: the buffers possess big capacity and the value of failure rates of workstations is identical. This assumption does not change the common results in differences of the productivity rates for the systems of serial structures. The initial data of manufacturing systems of serial structures are represented in Tables 4.12. Substituting initial data into equations of productivity rates of manufacturing systems of serial structure is defined by the values of productivity rates (Table 4.13).

Figure 4.18 illustrates the results of productivity rate for manufacturing systems of serial arrangement, where abscissa represents the

Table 4.12 Technological and technical data of manufacturing system of serial arrangement

Title	Data
Machining time, t_{mo} (min)	10.0
Auxiliary time, t_a (min)	0.3
Failure rate of workstation, λ_s	1.25×10^{-2}
Failure rate of transport system, λ_{tr}	4.0×10^{-4}
Failure rate of control system, $\lambda_{c.s}$	6×10^{-8}
Failure rate of buffer, λ_{bf}	6×10^{-8}
Mean repair time, m_r (min)	2.5
Number of workstations, q_s	10
Number of sections, n	3

Table 4.13 Productivity rate of manufacturing systems of serial structure

No	Manufacturing system	Equation of productivity rate
1	Independent workstations	$Q=\dfrac{1}{f_c(t_{mo}/q)+t_a}\times\dfrac{1}{1+m_r\lambda_{sb}}$
2	Automated line	$Q=\dfrac{1}{f_c(t_{mo}/q)+t_a}\times\dfrac{1}{1+m_r(\sum\limits_{i=1}^{q}\lambda_{s.i}+\lambda_{tr}+\lambda_{cs})}$
3	Automated line with sections and buffers	$Q_3=\dfrac{1}{f_c(t_{mo}/q)+t_a}\times\dfrac{1}{1+m_r[f_s\sum\limits_{i=1}^{q}\lambda_{s.i}/n+\lambda_b+\lambda_{tr}+\lambda_{cs}]}$
4	Automated line with buffers after each workstation	$Q_3=\dfrac{1}{f_c(t_{mo}/q)+t_a}\times\dfrac{1}{1+m_r(\lambda_{s.b}+\lambda_b+\lambda_{tr}+\lambda_{cs})}$

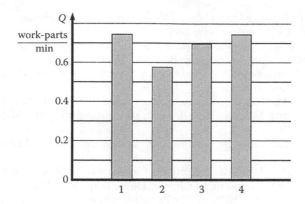

Figure 4.18 Productivity rate in histogram of manufacturing systems of serial arrangement.

following: 1 – manufacturing system of serial structure with an independent work of machines; 2 – an automated line of serial structure; 3 – an automated line of serial structure segmented on sections with buffers; 4 – an automated line of serial structure with buffers after each workstations. The values of productivity rate for manufacturing systems of serial structure are observed clearly through the histogram with bars.

The comparative analysis of the productivity rates for manufacturing systems of serial arrangement demonstrates the following. Highest productivity rate gives the manufacturing system of serial structure with an independent work of workstations (1) and lowest gives the automated line of serial structure with hard joining of workstations (2). The automated line segmented on sections with buffers enables increase in the productivity rate (3). Last design of the automated line with buffers after each workstation gives the productivity rate (4) a little less than the manufacturing systems of serial structure with an independent work of workstations. Nevertheless, all manufacturing systems are characterised by own properties. Systems number 1 and 4 occupy big production area, the latter one equipped with buffers and expensive. System number 2 is compact and comfortable for service, but more expensive than system 1 etc. Economists should solve which manufacturing design is preferable for production systems.

Bibliography

1. Altintas, Y. 2012. *Manufacturing Automation*. 2nd ed. Cambridge University Press. London.
2. Badiru, A.B., and Omitaomu, O.A. 2011. *Handbook of Industrial Engineering Equations, Formulas, and Calculations*. Taylor & Francis. New York.

3. Benhabib, B. 2005. *Manufacturing: Design, Production, Automation, and Integration.* 1st ed. Taylor & Francis. New York.
4. Ben-Daya, M., Duffuaa, S., and Raouf, A. 2000. *Maintenance, Modeling, and Optimization.* Springer Science+Business Media, New York.
5. Chryssolouris, G. 2006. *Manufacturing Systems: Theory and Practice.* 2nd ed. Springer. New York.
6. Groover, M.P. 2013. *Fundamentals of Modern Manufacturing: Materials, Processes, and Systems.* 5th ed. (Lehigh University). John Wiley & Sons. Hoboken, NJ.
7. Nof, S.Y. 2009. *Springer Handbook of Automation.* Purdue University. New York.
8. Rao, R.V. 2011. *Advanced Modeling and Optimization of Manufacturing Processes.* 1st ed. Springer, London, Dordrecht, Heidelberg, New York.
9. Shaumian, G.A. 1973. *Complex Automation of Production Processes.* Mashinostroenie. Moscow.
10. Shell, R.L., and Hall, E.L. 2000. *Handbook of Industrial Automation.* Marcel Dekker. Inc. New York.
11. Sin, T.C., Usubamatov, R., Fairuz, M.A. et al. 2015. Engineering mathematical analysis method for productivity rate in linear arrangement serial structure automated flow assembly line. *Mathematical Problems in Engineering.* 2015, Article ID 592061, 10 pages, DOI: 10.1155/2015/592061.
12. Usubamatov, R., Alwaise, A.M.A., and Zain, Z.M. 2013. Productivity and optimization of section-based automated lines of parallel-serial structure with embedded buffers. *The International Journal of Advanced Manufacturing Technology* 65(5). pp. 651–655.
13. Usubamatov, R., Sin, T.C., and Ahmad, R. 2016. Mathematical models for productivity of automated lines with different failure rates for stations and mechanisms. *The International Journal of Advanced Manufacturing Technology.* DOI 10.1007/s00170-015-7005-6.
14. Volchkevich, L.I. 2005. *Automation of Production Processes.* Mashinostroenie. Moscow.

chapter five

Manufacturing systems of parallel–serial arrangement

Production systems of parallel–serial arrangements represent combinations of parallel and serial manufacturing systems, which are designed according to the prescribed technological processes and specificity of work part designs. These systems are most productive in the industrial area, have a highest productivity rate that increased exponentially and depends on the number of parallel and serial workstations. The variations of parallel and serial arrangements depend on peculiarities of technological processes and the required productivity rates and restrictions in machine designs. Production systems of the parallel–serial structure can be arranged as manufacturing lines with several constructions. It can be a fully automated line with hard joining of parallel–serial structures, automated lines of serial structure and independent work of parallel one, automated lines of parallel structure and independent work of serial one, manufacturing lines of parallel–serial structures with independent work of all workstations or machine tools and other combinations. Which one is preferable depends on the attributes of economic efficiency of structural variants for manufacturing systems linked with problems of increasing the design complexity of systems, transportation of work parts to be machined, control systems, level of automation, etc. Each of the manufacturing structure is described by own properties, area of application and by mathematical models of productivity rates.

5.1   Introduction

Manufacturing systems of a parallel–serial arrangement represent a combination of parallel and serial manufacturing systems, which are laid out according to the prescribed technological processes. Manufacturing systems of a parallel–serial arrangement are most productive in the industrial area and have the highest productivity rate. Practically, manufacturers of different industries produce manufacturing systems with different number of parallel p and serial q machine tools or workstations. The arrangements represent peculiarities of technological processes and restrictions in machine designs. Manufacturing processes of the work parts with complex design is implemented by complex technologies, which are decomposed on several short consecutive operations. Each operation is implemented on one machine tool or workstations that form the manufacturing line of serial structure. In case the serial manufacturing line cannot provide the necessary level of productivity rate, it is added to parallel manufacturing serial lines that form manufacturing systems of the parallel–serial structure. Such method allows to dramatically increase the productivity rate of a manufacturing system in two ways, particularly, by segmentation and balancing of technological process and use of the same parallel processes. It means that productivity rate of manufacturing systems with parallel–serial arrangement can be increased exponentially.

Practically, manufacturing systems of the parallel–serial structure can be arranged as manufacturing or automated lines of parallel–serial structure with several constructions. Which one is preferable depends on the attributes of economic efficiency of structural and constructional variants for manufacturing systems. The design process of the systems of parallel–serial structures are linked with problems of increasing the complexity and cost of production systems, design the means for transportation of part machined, embedding the control systems, dispatching the work parts, increasing or decreasing the number of employees that depends on level of automation, etc.

Manufacturing systems of the parallel–serial structure can be arranged by the following variants.

- The system based on independent machine tools or workstations in parallel and serial arrangements.
- The system based on independent automated lines of parallel structure and serial arrangement.
- The system based on independent automated lines of serial structure and parallel arrangement.
- The system based on automated lines of parallel–serial structure with hard mechanical joining of all workstations, mechanisms and units.
- The system based on automated lines of parallel–serial structure segmented on sections with embedded buffers.
- The system based on automated lines of parallel structure and embedded buffers after each parallel line and serial arrangement.
- The system based on automated lines of serial structure segmented on sections with embedded buffers and parallel arrangement.
- The system based on rotor-type automated lines of parallel–serial arrangements and combinations of parallel–serial structures.
- Other structural combinations.

Each of these manufacturing structures can be described by the properties that represented the earlier chapters for separate parallel and serial manufacturing and automated lines. For the following analysis, properties and area of application for different types of manufacturing systems are used

to form the systems of parallel–serial arrangements. However, it is necessary to underline that new systems of parallel–serial arrangements combine specific properties of separate manufacturing systems and combine
mathematical models of productivity rates.

Modern manufacturing systems with complex constructions require
reliable mathematical models to calculate the manufacturing system's
productivity rate. The productivity rate of a manufacturing system depends
on several components, namely, segmentation and balancing of a technological process, reliability of machines and managerial and organisational
problems. The productivity is increased in powered proportion to the
number of parallel and serial stations. Increasing the number of stations in
the automated line leads to a proportional increase of its downtime due to
the reliability of a complex system. Solving the problem of the automated
lines' optimal structure with complex design by the criterion of maximal
productivity and efficiency is a crucial issue. Detailed considerations of
mathematical models for each type of parallel–serial manufacturing systems and automated lines are presented later. For simplicity of presentation
and to avoid cumbersome expressions for the equations of productivity
rate, the assumption that the failure rates of serial workstations are equal
at the same workstations of parallel arrangement is considered. However,
if necessary to have corrected equations for workstations with different
attributes of reliability, it is necessary to refer the mathematical models
that represented in the corresponding earlier chapters.

5.1.1   Productivity rate of manufacturing systems with parallel–serial arrangement and independent work of workstations

The symbolic picture of the manufacturing systems with parallel–serial
arrangement and an independent work of workstations is represented
in Figure 5.1. This parallel–serial arrangement is characterised by several negative and positive properties according to the number of parallel
and serial workstations. The properties of such line are described by the
combined properties of the manufacturing system of parallel and serial
structures represented in Chapters 3 and 4. Failure of any machine tool or
workstation in the manufacturing system does not lead to a downtime of
other workstations that continue working. The lack of such arrangement
of manufacturing line is occupation of a big production area compared
with compact designs of the automated lines.

Mathematical model of productivity rate for manufacturing system
of the parallel–serial arrangement with independent work of q serial and
p parallel workstations is represented by combined Eqs. (3.1) and (4.2) for
parallel and serial structures of manufacturing systems. The equation

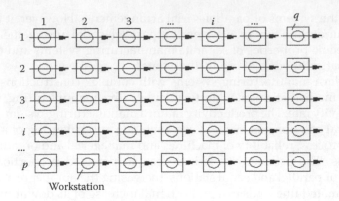

Workstation

Figure 5.1 Manufacturing system of the parallel–serial arrangement with independent work of q serial and p parallel workstations.

for productivity rate for a manufacturing system of the parallel–serial arrangement is represented by the following expression:

$$Q = \frac{p}{\dfrac{t_{mo}}{q} f_c + t_a} \times \frac{1}{1 + m_r \lambda_{sb}} \tag{5.1}$$

where all parameters are as specified earlier.

Equation (5.1) contains the component of availability $A = \dfrac{1}{1 + m_r \lambda_{sb}}$ that is considered only for one workstation, because the productivity rate of manufacturing system of parallel–serial arrangement is calculated for one bottleneck machine tool or workstation. Other parallel workstations are implementing the same process and represented in Eq. (5.1) as the number p work parts machined by p_s parallel workstations. Analysis of Eq. (5.1) enables to depict the change in productivity rate of the manufacturing system of parallel–serial arrangement with increasing number of parallel and serial machine tools or workstations. This change is represented in Figure 5.2 and demonstrates increase in the productivity rate for a manufacturing system of parallel–serial arrangement versus the number of independent workstations.

Equation (5.1) demonstrates that the productivity rate of manufacturing systems of parallel–serial arrangement increased hyperbolically according to the numbers of parallel and serial workstations.

A working example

A manufacturing system of parallel–serial arrangement with independent workstations is designed with the technical and timing data referred to one product and is shown in Table 5.1. The productivity rate per cycle time, total productivity rates and availability for the given

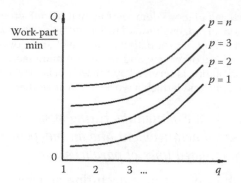

Figure 5.2 Productivity rate of the manufacturing systems of parallel–serial arrangement with independent number of parallel p and serial q workstations.

Table 5.1 Technical data of a manufacturing system

Title	Data
Total machining time, t_{mo} (min)	12.0
Correction factor for a bottleneck workstation, f_c	1.3
Auxiliary time, t_a (min)	0.3
Failure rate of the workstation, λ_{sb}	0.02
Mean repair time, m_r (min)	3.0
Number of parallel workstations, p_s	5
Number of serial workstations, q	10

data of the manufacturing system of parallel–serial arrangement are calculated.

Solution: Substituting the initial data of Table 5.1 into Eq. (5.1) and calculating yield the following results:

The productivity rate per cycle time is as follows:

$$Q = \frac{p}{\dfrac{t_{mo}}{q} f_c + t_a} = \frac{5}{\dfrac{12}{10} \times 1.3 + 0.3} = 2.688 \text{ work parts/min}$$

The availability is as follows:

$$A = \frac{1}{1 + m_r \lambda_{sb}} = \frac{1}{1 + 3.0 \times 0.02} = 0.943$$

The total productivity rate of the manufacturing line of parallel–serial arrangement is as follows:

$$Q = \frac{p}{\dfrac{t_{mo}}{q} f_c + t_a} \times \frac{1}{1 + m_r \lambda_{sb}} = 2.688 \times 0.943 = 2.535 \text{ work part/min}$$

Conducted calculations demonstrate that manufacturing systems of parallel–serial arrangement with independent work of workstations do not have the limit of productivity. Failure of any workstation does not lead to stop the other one and the entire manufacturing system. The failed workstation is serviced by the technicians who remove failures per short time and launch a workstation at workable condition.

5.1.2 Productivity rate of manufacturing systems of parallel–serial arrangement and an independent work of automated lines of parallel structure

The symbolic picture of the manufacturing systems of parallel–serial arrangement and independent work of automated lines of parallel structure is represented in Figure 5.3. The properties of such lines are described by the combined properties of an automated line of parallel structure and the manufacturing system of serial arrangement represented in Chapters 3 and 4. The failure of any workstation in the automated line of parallel structure leads to a downtime of one line, but other serial automated lines of parallel structure go on working. The manufacturing system of parallel–serial arrangement occupies less production area, because automated lines of parallel structures are compact in design.

Combined Eqs. (3.2) and (4.2) for parallel and serial structures of a manufacturing line present the mathematical model for productivity rate of manufacturing system of the parallel–serial structure with an independent work of q parallel automated lines with p parallel workstations. New equation is represented by the following expression:

$$Q = \frac{p}{\frac{t_{mo}}{q} f_c + t_a} \times \frac{1}{1 + m_r \left(p_s \lambda_{sb} + \lambda_{cs} + \lambda_{tr} \right)} \tag{5.2}$$

where all parameters are as specified earlier.

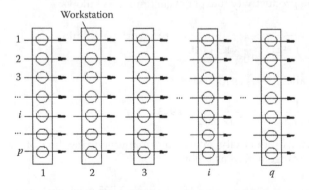

Figure 5.3 Manufacturing system of the parallel–serial arrangement with an independent work of q parallel automated lines with p parallel workstations.

Equation (5.2) demonstrates the increase in the number of parallel workstations p and serial workstations q leading to increase in the productivity rate per cycle time. The value of the workstation's failure rates is increasing on p_s times that equals to the number of parallel workstations. The availability for the manufacturing system of parallel–serial arrangement with automated lines of parallel structure is decreasing. Figure 5.4 depicts the change in value of components of Eq. (5.2) versus the change in the number of parallel p_s and serial workstations q.

Preliminary analysis of Eq. (5.2) demonstrates that there is maximal productivity rate for the manufacturing system of parallel–serial arrangement with serial automated lines of parallel structure. This maximal productivity rate Q_{max} is defined by solving the mathematical limit of Eq. (5.2).

$$\lim_{p \to \infty} Q = \lim_{p \to \infty} \left(\frac{p}{\dfrac{t_{mo}}{q} f_c + t_a} \times \frac{1}{1 + m_r \left(p_s \lambda_{sb} + \lambda_{cs} + \lambda_{tr} \right)} \right) = \frac{1}{\left(\dfrac{t_{mo}}{q} f_c + t_a \right) m_r \lambda_{sb}}$$

(5.3)

where all components are as specified earlier.

Analysis of Eq. (5.3) demonstrates the maximal productivity rate for the manufacturing system of parallel–serial arrangement based on the serial automated lines of parallel structure changes with the number of serial workstations and reliability indices of one workstation. Figure 5.4 depicted by Eqs. (5.2) and (5.3) demonstrates that the productivity rate for the manufacturing system of parallel–serial arrangement is increased

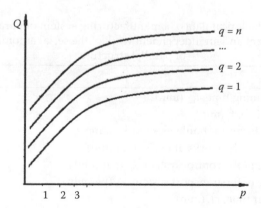

Figure 5.4 Productivity rate of manufacturing systems of parallel–serial arrangement and an independent work of automated lines of parallel structure.

asymptotically with the number of serial automated lines of parallel structure and reached a defined limit.

The manufacturing system of serial arrangement has a defined limit in productivity. However, automated lines of parallel structure arranged in serial lines of independent work enable the growth of productivity rate for the manufacturing systems of parallel–serial structure with increasing number of parallel-automated lines.

A working example

A manufacturing line of parallel–serial arrangement with an independent work of the serial automated lines of parallel structure is designed with the technical and timing data referred to one product that is shown in Table 5.2.

The productivity rate per cycle time, total productivity rate and availability for the given data of the manufacturing system of parallel–serial arrangement are calculated.

Solution: Substituting the initial data of Table 5.2 into Eq. (5.2) and calculating lead to the following results.

The productivity rate per cycle time of the manufacturing system is as follows:

$$Q = \frac{p}{\frac{t_{mo}}{q} f_c + t_a} = \frac{8}{\frac{26}{12} \times 1.2 + 0.3} = 3.2 \text{ work part/min}$$

The availability the manufacturing system is as follows:

$$A = \frac{1}{1 + m_r \left(p_s \lambda_s + \lambda_{cs} + \lambda_{tr} \right)} = \frac{1}{1 + 3.0(8 \times 0.025 + 0.005 + 0.0005)} = 0.581$$

Table 5.2 Technical data of a manufacturing system of parallel–serial arrangement and an independent work of the serial automated lines of parallel structure

Title	Data
Total machining time, t_{mo} (min)	26.0
Auxiliary time, t_a (min)	0.3
Correction factor for bottleneck workstation, f_c	1.2
Failure rate of the workstation, λ_{sb} (per min)	0.025
Failure rate of the control system, λ_{cs} (per min)	0.005
Failure rate of the transport system, λ_{tr} (per min)	0.0005
Mean repair time, m_r (min)	3.0
Number of parallel stations in automated line, p_s	8
Number of serial workstations, q	12

The total productivity of the manufacturing system with independent automated line of parallel structure arranged in the line of serial structure is as follows:

$$Q_1 = \frac{p}{\frac{t_{mo}}{q}f_c + t_a} \times \frac{1}{1 + m_r\left(p_s\lambda_s + \lambda_{cs} + \lambda_{tr}\right)} = 3.2 \times 0.581 = 1.86 \text{ work part/min}$$

Conducted calculations demonstrate technical and technological attributes of the manufacturing system of parallel–serial arrangement with independent serial automated lines of parallel structure.

5.1.3   Productivity rate of manufacturing systems of parallel–serial arrangement and an independent work of automated lines of serial structure

The symbolic picture of the manufacturing systems of parallel–serial arrangement and an independent work of automated lines of serial structure is represented in Figure 5.5. The properties of such line are described by the properties of the manufacturing system of parallel structure and the automated line of serial structure presented in Chapters 3 and 4. The failure of any workstation in the automated line of serial structure leads to downtime of this line, but other automated serial lines go on working. The properties of the manufacturing line of parallel–serial arrangement are described by properties of the automated lines of serial and parallel structure.

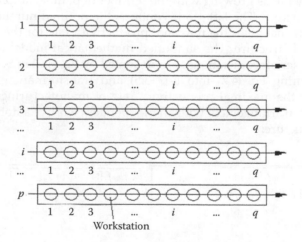

Figure 5.5 Manufacturing system of the parallel–serial arrangement with an independent work of parallel-automated lines p and serial structure with q workstations.

Combined Eqs. (3.1) and (4.4) represent a mathematical model for productivity rate of manufacturing system of the parallel–serial structure arrangement with an independent work of p automated lines and the serial structure with q workstations. The equation for productivity rate for such manufacturing system is represented by the following expression:

$$Q = \frac{p}{\dfrac{t_{mo}}{q} f_c + t_a} \times \frac{1}{1 + m_r \left(\lambda_{sav} \displaystyle\sum_{i=1}^{q} f_{s.i} + \lambda_{tr} + \lambda_{cs} \right)} \quad (5.4)$$

For an ideal manufacturing system with equal failure rates of serial stations:

$$Q = \frac{p}{\dfrac{t_{mo}}{q} f_c + t_a} \times \frac{1}{1 + m_r (q\lambda_s + \lambda_{tr} + \lambda_{cs})} \quad (5.5)$$

where all parameters are as specified earlier.

Equation (5.4) demonstrates that increase in the number of parallel workstations p and serial workstations q leads to an increase in the productivity rate per cycle time. The increase in the number of serial workstations leads to increase in the failure rate of the automated line of serial structure. Hence, the value of availability for the manufacturing system of parallel–serial arrangement is represented by an automated line of serial structure.

Preliminary analysis of Eq. (5.4) demonstrates that the maximal of productivity rate is growing with the number of parallel automated lines of serial structure. The latter one contains the optimal number of serial workstations that gives the maximal productivity rate for the automated line of serial structure. The simplified mathematical model for optimal numbers of serial workstations is represented by Eq. (4.7) in Section 4.2.

Substituting Eq. (4.7) into Eq. (5.5) and transformation yields the equation of the maximal productivity rate of manufacturing system of parallel–serial arrangement with an independent work of automated lines of serial structure:

$$Q_{max} = \frac{p}{\sqrt{\dfrac{t_{mo} f_c t_a \lambda_s}{(1/m_r) + \lambda_{cs} + \lambda_{tr}}} + t_a} \times \frac{1}{1 + m_r \left\{ \sqrt{\dfrac{\lambda_s t_{mo} f_c [(1/m_r) + \lambda_{cs} + \lambda_{tr}]}{t_a}} + \lambda_{tr} + \lambda_{cs} \right\}} \quad (5.6)$$

where all parameters are as specified earlier.

Figure 5.6 depicted by Eqs. (4.4) and (4.5) demonstrates that increase in the productivity rate of the manufacturing system of parallel–serial

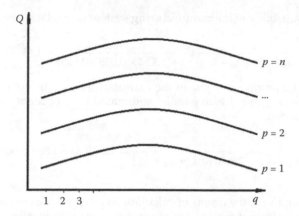

Figure 5.6 Productivity rate of manufacturing systems of parallel–serial arrangement and an independent work of automated lines of serial structure.

arrangement with an independent work of automated lines of serial structure increases with the number of parallel automated lines.

Figure 5.6 demonstrates that the automated line of serial structure has maximal productivity rate with optimal number of serial workstations. Increasing the number of the parallel lines gives the main growth of the productivity rate.

A working example

A manufacturing line of parallel–serial arrangement with an independent work of parallel automated lines of serial structure is designed with the technical and timing data referred to one product and is shown in Table 5.2 and is represented in Section 5.1.2. The productivity rate per cycle time, total productivity rate and availability for the given data of the manufacturing system of parallel–serial arrangement with automated serial lines are calculated.

Solution: Substituting the initial data of Table 5.2 into Eq. (5.5), the calculating results are represented by the following.

The optimal number of workstations that gives maximal productivity rate for the automated line of serial structure is as follows:

$$q_{opt} = \sqrt{\frac{t_{mo} f_c [(1/m_r) + \lambda_{cs} + \lambda_{tr}]}{t_a \lambda_s}} = \sqrt{\frac{26.0 \times 1.2 \times [(1/3.0) + 0.005 + 0.0005]}{0.3 \times 0.025}} = 37$$

The productivity rate per cycle time of the manufacturing system is as follows:

$$Q = \frac{p}{\dfrac{t_{mo}}{q} f_c + t_n} = \frac{8}{\dfrac{26}{12} \times 1.2 + 0.3} = 3.2 \text{ work part/min}$$

The availability of the manufacturing system is as follows:

$$A = \frac{1}{1 + m_r \left(q_s \lambda_s + \lambda_{cs} + \lambda_{tr} \right)} = \frac{1}{1 + 3.0(12 \times 0.025 + 0.005)} = 0.48$$

The total productivity rate of the manufacturing system of parallel–serial arrangement with parallel automated lines of serial structure is as follows:

$$Q_2 = \frac{p}{\dfrac{t_{mo}}{q} f_c + t_a} \times \frac{1}{1 + m_r \left(q_s \lambda_s + \lambda_{cs} + \lambda_{tr} \right)} = 3.2 \times 0.48 = 1.536 \text{ product/min}$$

Comparatively, the results of calculations for two types of manufacturing systems of parallel–serial arrangements with identical technical parameters, presented in Sections 5.1.2 and 5.1.3, with automated lines of parallel and serial structures demonstrate that their productivity rates are different. The serial automated lines of a parallel structure give higher productivity rate Q_1 than the parallel automated lines of serial structure Q_2, i.e. $Q_1 > Q_2$, or $1.86 > 1.536$ product/min. This result yields the following conclusion. In case of the design of the manufacturing system of parallel–serial arrangement with identical technical data and restriction in automation for the automated lines of parallel or serial structure, the first one is preferable by criterion of maximal productivity rate.

5.1.4 Productivity rate of manufacturing systems of parallel–serial arrangement and an independent work of automated lines of serial structure segmented on sections with buffers

The symbolic picture of the manufacturing systems of parallel–serial arrangement and an independent work of automated lines of serial structure segmented on sections with buffers is represented in Figure 5.6. The properties of such line are described by the properties of the manufacturing system of parallel structure and the automated line of serial structure presented in Chapters 3 and 4. The failure of any workstation in the automated line of serial structure leads to a downtime of this line only, but other automated serial lines go on working. The properties of the manufacturing line of parallel–serial arrangement are described by properties of the automated lines of serial and parallel structure.

The equation of the productivity rate for the considered manufacturing line is represented by combination of equations for the automated lines of parallel structure (Eq. 3.1) and serial one (Eqs. 4.16 and 4.17) segmented on sections with buffers of big or limited capacities. Combined Eqs. (3.1) and (4.16) or (4.17) represent a mathematical model for productivity rate of manufacturing system of the parallel–serial structure arrangement with an independent work of automated lines of the serial structure with buffers. The equations for productivity rate of such manufacturing systems are represented by the following expression:

a. Manufacturing system with buffers of limited capacity

$$Q = \frac{p}{\dfrac{t_{mo}}{q} f_c + t_a} \times \frac{1}{1 + m_r \left[\dfrac{f_s \displaystyle\sum_{i=1}^{q} \lambda_{s.i}}{n} + \Delta\lambda_{f(i-k)} + \cdots + \Delta\lambda_{f(i-2)} + \Delta\lambda_{f(i-1)} + \Delta\lambda_{b(i+1)} + \Delta\lambda_{b(i+2)} + \cdots + \Delta\lambda_{b(i+n)} + \lambda_{bf} + \lambda_c + \lambda_{tr} \right]}$$

(5.7)

b. Manufacturing system with buffers of big capacity

$$Q = \frac{p}{\dfrac{t_{mo}}{q} f_c + t_a} \times \frac{1}{1 + m_r \left[\dfrac{f_s \displaystyle\sum_{i=1}^{q} \lambda_{s.i}}{n} + \lambda_{bf} + \lambda_c + \lambda_{tr} \right]}$$

(5.8)

where all parameters are as specified earlier.

Analysis of Eqs. (5.7) and (5.8) demonstrates increase in the number of parallel lines of serial structure leading to increase in the productivity rate of the manufacturing system. The maximal productivity rate is growing with the number of parallel automated lines of serial structure with buffers. Buffered automated lines of serial structure for the given number of section contain an optimal number of serial workstations that gives the maximal productivity rate.

Figure 5.7 depicted by Eq. (5.7) or (5.8) demonstrates that increasing the productivity rate for the manufacturing system of parallel–serial arrangement with an independent work of automated lines of serial structure is increased with the number p of parallel automated lines and the number of the sections n in the serial buffered line.

Figure 5.8 demonstrates the manufacturing systems with parallel and independent automated lines of serial structure with buffers give the main growth of the productivity rate with increase in the number of parallel lines.

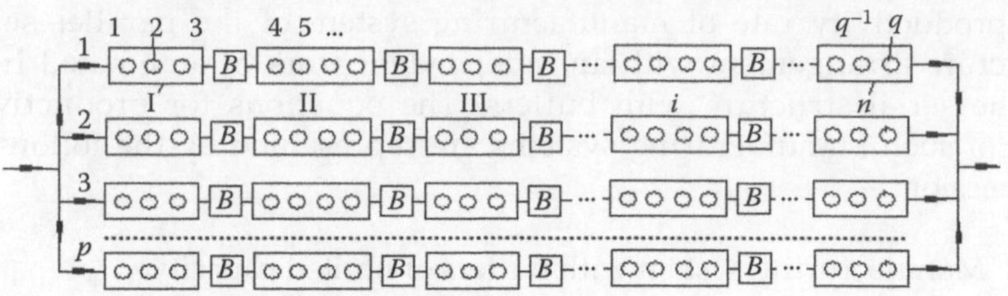

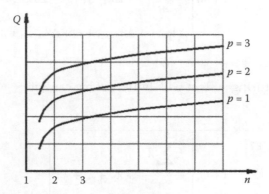

Figure 5.7 Manufacturing system of the parallel–serial arrangement with an independent work of the automated lines of serial structure with q workstations segmented on sections n with buffers B and parallel lines p.

Figure 5.8 Productivity rate of manufacturing systems of parallel–serial arrangement and an independent work of automated lines of serial structure.

5.2 *Productivity rate of an automated line of linear arrangement and parallel–serial structure*

Practically, manufacturers of different industries produce automated lines with a different number of parallel p and serial q stations. The structural designs of automated lines connected with peculiarities of technological processes and technical restrictions. Two types of manufacturing systems with complex structures represent the design of automated lines with serial–parallel structure. The first type of an automated line of serial–parallel structure has a linear arrangement, which is applied for processes having long machining times like machining of housing type work parts, shafts with complex design, gears and so on. The second type of an automated line has a rotor-type arrangement, whose application is found in various branches of the industry with short machining processes like pressing, coining, filling by liquids the bottles and cans and so forth. The mathematical models for the productivity rate of these two types of machines are different due to differences in design.

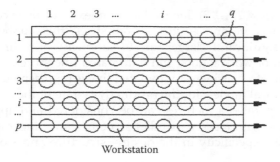

Figure 5.9 Automated line of the parallel–serial structure with p parallel lines and q serial workstations.

The symbolic picture of the automated line of parallel–serial structure and linear arrangement is represented in Figure 5.9. This automated production line can be considered as a system for the collection of serial and parallel workstations arranged according to a certain structure that depends on a technological process of machining work parts. The properties of such line are described by the combined properties of the automated lines of parallel and serial structure presented in Chapters 3 and 4. The automated line of parallel–serial structure and linear arrangement is designed with hard connections between all workstations of parallel and serial location. The failure of any workstation in such designed line leads to a downtime of the entire line until completion of failure fixing. The specification and peculiarities of the design for the automated lines of parallel–serial structure are represented by the following properties:

- All parallel and serial workstations start to work at one time simultaneously and after finishing previous operations.
- Any failure, in either serial or parallel workstations or other units and mechanisms, leads to stoppage of an entire automated line, due to the mechanical hard joining of all mechanisms.
- The control and transport systems serve all serial and parallel workstations
- All work parts carry forward the linear design of the transport mechanism to the following parallel–serial workstations.

This automated line of parallel–serial structure is characterised by the negative and positive properties of the automated lines according to the number of parallel and serial workstations. Practically, in manufacturing area, machining the work parts on the automated lines of parallel–serial structures with linear arrangement is designed only with two to three parallel lines ($p_s = 2, ..., 3$) due to the complexity of their constructions. However, in electric and electronic industries, the number of parallel lines can be significantly larger due to simplicity of technological processes, design of workstations and automated line structures. The schematic diagram of a considered automated line is represented in Figure 5.9. The mathematical models for the productivity rate of the automated line of serial and parallel structures can be considered and applied for the automated line with a serial–parallel structure and linear arrangement, which has own specificity in the design, structure and work described earlier.

Automated line of the parallel–serial structure is designed with p parallel and q serial workstations, which are the components of the productivity rate equation. Based on mathematical considerations of failure rates for automated lines of parallel and serial structures considered earlier, it is possible to represent the equations for the productivity rate and the availability of automated lines with serial–parallel structure. Equations (3.2) and (4.5) express mathematical model of productivity rate for automated line of the parallel–serial structure with p parallel automated lines of q serial workstations. The equation for a considered structure of an automated line is represented by the following expression:

$$Q = \frac{p}{\frac{t_{mo}}{q} f_c + t_a} \times \frac{1}{1 + m_r \left(p_s \sum_{i=1}^{q} \lambda_{s.i} + \lambda_{tr} + \lambda_{cs} \right)} \qquad (5.9)$$

The simplified mathematical model for productivity rate of the automated line accepts equal failure rates of workstations and does not give accurate solution and is represented by the following equation [9,15]:

$$Q = \frac{p}{\frac{t_{mo}}{q} f_c + t_a} \times \frac{1}{1 + m_r (p_s q \lambda_s + \lambda_{tr} + \lambda_{cs})} \qquad (5.10)$$

where all parameters are as specified earlier.

Mathematical models demonstrate exponential increasing the productivity rate per cycle time, which is the first factor of Eqs. (5.8)–(5.10) and an exponential decrease in the failure rates represented by the product of numbers of parallel and serial stations at the second factor, ($p_s q$), which is availability of the automated line of parallel–serial structure.

Equation (5.10) of productivity rate, based on the average failure rates of the automated lines workstations, is effectively used for the analytical optimisation analysis of the automated line parameters. Equation (5.10) contains two variables, i.e. the number of q serial and p

parallel workstations. Analysis of Eq. (5.10) gives extreme of the function, i.e. there is maximal productivity rate for the defined number of serial workstations. This maximum is defined by solving the first partial derivative of Eq. (5.10) with the variable of q workstations, when other parameters are constant.

$$\frac{\partial Q}{\partial q} = \frac{\partial \left[\dfrac{p}{\dfrac{t_{mo}}{q}f_c + t_a} \times \dfrac{1}{1 + m_r(p_s q \lambda_s + \lambda_{cs} + \lambda_{tr})} \right]}{\partial q} = 0,$$

giving rise to the following:

$$p\left\{ -\frac{t_{mo}}{q^2}f_c[1 + m_r(\lambda_{cs} + \lambda_{tr})] + p_s t_a \lambda_s m_r \right\} = 0,$$

solving this equation yields the following expression

$$q_{opt} = \sqrt{\frac{t_{mo}f_c[(1/m_r) + \lambda_{cs} + \lambda_{tr}]}{p_s t_a \lambda_s}} \tag{5.11}$$

where all parameters are as specified earlier.

Similar optimal number of serial workstations gives Eqs. (5.8) and (5.9) that is solved by numerical method. Equation (5.11) represents the optimal number of serial workstations, which gives the maximal productivity rate for the automated line of the serial–parallel structure with the given number of parallel workstations. Optimal number of serial workstation in the automated line depends on the number of parallel workstations, reliability parameters of workstation, transport and control systems and bottleneck machining and auxiliary times. In addition, the optimal number of serial workstations decreased with the increase in the number of parallel workstations. Substituting Eq. (5.11) into Eq. (5.10) and transformation yield the equation of the maximal productivity for the automated line of parallel–serial structure:

$$Q_{max} = \frac{p}{\sqrt{\dfrac{p_s t_{mo} f_c t_a \lambda_s}{(1/m_r) + \lambda_{cs} + \lambda_{tr}}} + t_a}$$

$$\times \frac{1}{1 + m_r\left\{ \sqrt{\dfrac{p_s \lambda_s t_{mo} f_c[(1/m_r) + \lambda_{cs} + \lambda_{tr}]}{t_a}} + \lambda_{tr} + \lambda_{cs} \right\}} \tag{5.12}$$

Solving the first partial derivative of Eq. (5.10) with variable by the second parameter p_s, when other parameters are constant, gives the following:

$$\frac{\partial Q}{\partial p} = \frac{\partial \left[\dfrac{p}{\dfrac{t_{mo}}{q} f_c + t_a} \times \dfrac{1}{1 + m_r(p_s q \lambda_s + \lambda_{cs} + \lambda_{tr})} \right]}{\partial p} = 0,$$

giving rise to the following

$$\frac{1}{\left(\dfrac{t_{mo}}{q} f_c + t_a \right)} \times \frac{1}{1 + m_r(p_s q \lambda_s + \lambda_{cs} + \lambda_{tr})} + \frac{p}{\left(\dfrac{t_{mo}}{q} f_c + t_a \right)}$$

$$\times \frac{-m_r q \lambda_s}{\left[1 + m_r(p_s q \lambda_s + \lambda_{cs} + \lambda_{tr}) \right]^2} = 0$$

Solving this equation and transformation yield the following result: $1 + m_r(\lambda_{cs} + \lambda_{tr}) = 0$. This result shows that the automated line of parallel–serial structure does not have an optimal number of the parallel workstations that gives the maximal productivity rate.

Figure 5.10 depicted by Eqs. (5.8), (5.9) and (5.10) demonstrates some interesting and important dependency of parameters in productivity rate. The maximal productivity rate is increasing with the number of parallel workstations p_s, which leads to a decrease in the number of serial workstations q according to the equation of productivity rate, Eq. (5.9). For the case with many parallel workstations, the curves of Eqs. (5.8) and (5.9) can have drastic form and to define the optimal number of serial workstations should be accepted very carefully. It is preferable to accept the number of optimal serial workstations by the curve of productivity rate on the slope to the right of the optimum point, because it is more gradual compared with the left side, thus giving less error. In the case of small number of parallel workstations, the optimal number of serial workstations locates on the curve with a light incline. This form of curve enables decreasing the optimal number of serial stations without a sensitive drop in the productivity rate. In such case, the final decision regarding optimal number of workstations depends on economic considerations.

The diagrams in Figure 5.8 demonstrate by dotted line the tendency that in case of maximal productivity increase in the number of parallel workstations leads to a decrease in the number of serial workstations. This

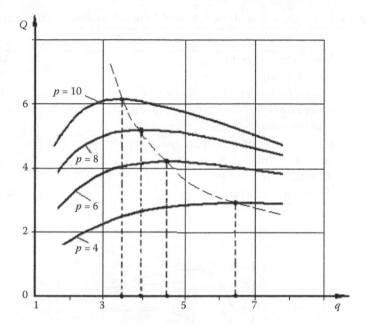

Figure 5.10 Changes in the productivity rate of an automated line of parallel–serial structure with changes in the number of serial q and parallel p workstations.

tendency is attributed only for automated lines of parallel–serial structure with different designs.

Defined equations, of the productivity rate of the automated line of parallel–serial structure, optimal number of serial workstations and maximal productivity rate, respectively, enable for the calculation of more trustworthy technical characteristics and their productivity rate as a function of the number of parallel and serial workstations. These equations enable for evaluation on the variations of productivity rate for an automated line with different parallel–serial structures.

A working example 1

An automated line of parallel–serial structure with hard connections of parallel and serial workstations arranged in linear system is designed with the technical and timing data referred to one product shown in Table 5.3. The optimal number of serial workstations that gives the maximal productivity rate is defined. Results represent in diagram the change in the productivity rate of the automated line versus the change in the number of parallel and serial workstations.

Solution: Substituting the initial data of Table 5.3 into Eq. (5.7) and calculating yield the results presented in Figure 5.11

Table 5.3 Technical data of the automated line of parallel–serial structure

Title	Data
Total machining time, t_{mo} (min)	16.0
Auxiliary time, t_a (min)	0.3
Correction factor for bottleneck workstation, f_c	1.2
Average failure rate of the workstation, λ_s (per min)	0.025
Failure rate of the control system, λ_{cs} (per min)	0.005
Failure rate of the transport system, λ_{tr} (per min)	0.0005
Mean repair time, m_r (min)	3.0
Number of parallel workstations, p_s	2,…,6
Number of serial workstations, q	4,…,28

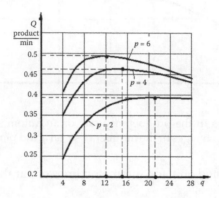

Figure 5.11 Changes in the productivity rate of an automated line of parallel–serial structure.

The optimal number of serial workstations q_{opt} that gives the maximal productivity rate Q_{max} is as follows:

- For the automated line of two parallel workstations

$$q_{opt} = \sqrt{\frac{t_{mo} f_c [(1/m_r) + \lambda_{cs} + \lambda_{tr}]}{p_s t_a \lambda_s}} = \sqrt{\frac{16.0 \times 1.2 \times [(1/3.0) + 0.005 + 0.0005]}{2 \times 0.3 \times 0.025}}$$

$$= 20.82$$

The number of stations is rounding $q_{opt} = 21$.

$$Q_{max} = \frac{p}{\dfrac{t_{mo}}{q_{opt}} f_c + t_a} \times \frac{1}{1 + m_r (p_s q_{opt} \lambda_s + \lambda_{tr} + \lambda_{cs})}$$

$$= \frac{2}{\dfrac{16.0}{21} \times 1.2 + 0.3} \times \frac{1}{1 + 3.0(2 \times 21 \times 0.025 + 0.005 + 0.0005)}$$

$$= 0.395 \text{ work part/min}$$

- For the automated line of four parallel stations

$$q_{opt} = \sqrt{\frac{t_{mo}f_c[(1/m_r)+\lambda_{cs}+\lambda_{tr}]}{p_s t_a \lambda_s}}$$

$$= \sqrt{\frac{16.0\times1.2\times[(1/3.0)+0.005+0.0005]}{4\times0.3\times0.025}} = 14.72$$

The number of workstations is rounding $q_{opt} = 15$.

$$Q_{max} = \frac{p}{\dfrac{t_{mo}}{q_{opt}}f_m + t_a} \times \frac{1}{1+m_r(p_s q_{opt}\lambda_s + \lambda_{tr} + \lambda_{cs})}$$

$$= \frac{4}{\dfrac{16.0}{15}\times1.2+0.3} \times \frac{1}{1+3.0(4\times15\times0.025+0.005+0.0005)}$$

$$= 0.458 \text{ work part/min}$$

- For the automated line of six parallel stations

$$q_{opt} = \sqrt{\frac{t_{mo}f_c[(1/m_r)+\lambda_{cs}+\lambda_{tr}]}{p_s t_a \lambda_s}}$$

$$= \sqrt{\frac{16.0\times1.2\times[(1/3.0)+0.005+0.0005]}{6\times0.3\times0.025}} = 12.02$$

The number of stations is rounding $q_{opt} = 12$.

$$Q_{max} = \frac{p}{\dfrac{t_{mo}}{q_{opt}}f_c + t_a} \times \frac{1}{1+m_r(p_s q_{opt}\lambda_s + \lambda_{tr} + \lambda_{cs})}$$

$$= \frac{6}{\dfrac{16.0}{12}\times1.2+0.3} \times \frac{1}{1+3.0(6\times12\times0.025+0.005+0.0005)}$$

$$= 0.492 \text{ work part/min}$$

Conducted calculations and Figure 5.11 confirm the particular attributes of the automated lines of parallel–serial structures of linear arrangement and the tendency in the change in the productivity rates with the change in the number of parallel and serial workstations.

A working example 2

The automated line with serial–parallel action can be designed with variants of $q = 5,...,15$ serial workstations and three parallel workstations (Figure 5.7). The basic technical and technological data for the automated line are presented in Tables 5.4 and 5.5. Substituting

Table 5.4 Technical data for an automated line of parallel–serial structure

Title	Data
Total machining time, t_{mo} (min)	15
Auxiliary time, t_a (min)	0.3
Number of serial workstations, q	2,…,14
Number of parallel workstations, p_s	3
Correction factor for machining time of the bottleneck workstation, f_c	1.2

Table 5.5 Reliability indices for an automated line of parallel–serial structure

Title	q	$\lambda_{s,i}$ (k/min)
Failure rate of the serial workstations q, λ_s	1–4	7.0×10^{-2}
	5–8	5.0×10^{-2}
	9–12	8.0×10^{-2}
	13–16	6.0×10^{-2}
	17–20	9.0×10^{-2}
	21–24	5.0×10^{-2}
		Average 6.666×10^{-2}
Failure rate of the control system, λ_c		8.0×10^{-3}
Failure rate of the transport system, λ_{tr}		5.0×10^{-4}
Mean repair time, $m_r = 3.0$ min		

the initial data (Tables 5.4 and 5.5) into Eqs. (5.8), (5.9) and calculating yield Eqs. (Q_1) and (Q_2). The calculations and results represented in Table 5.6 are used to depict the diagram of changes in the productivity rate versus the number of serial and parallel stations with different failure rates (Figure 5.12).

$$Q_1 = \frac{p}{\frac{15}{q} \times 1.2 + 0.3} \times \frac{1}{1 + 3\left(3\sum_{i=1}^{q} \lambda_{s.i} + 8 \times 10^{-3} + 5 \times 10^{-4}\right)}$$

$$Q_2 = \frac{p}{\frac{15}{q} \times 1.2 + 0.3} \times \frac{1}{1 + 3(3q6.666 \times 10^{-2} + 8 \times 10^{-3} + 5 \times 10^{-4})}$$

Table 5.6 Productivity rate for an automated line of parallel–serial structure with q serial and p parallel workstations

$p = 3, q$	4	8	12	16	20	24
Q, work part/min	0.173	0.22	0.202	0.202	0.183	0.185
Q_{av}, work part/min	0.182	0.201	0.202	0.198	0.192	0.185

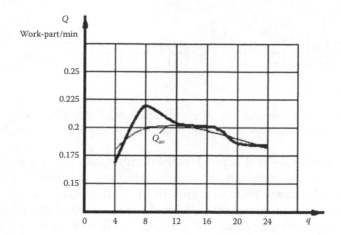

Figure 5.12 Productivity rate for an automated line of parallel–serial structure versus the number of serial and parallel workstations with different failure rates.

The difference in the results calculated by Eqs. (5.8) and (5.9) that represented in Table 5.6 and Figure 5.12 is considerable. The simplified approach in the calculation produces serious deviation that cannot be accepted for the final solutions [9,15].

5.2.1 Productivity rate of an automated line of parallel–serial structure segmented on sections with buffers

In engineering, the automated production lines of parallel–serial structures segmented on sections with buffers are most complex in design. The mathematical models of the productivity rate for these complex structure automated lines are also complex and represented as a function of the number of serial sections with buffers and workstations, parallel–serial structures and technical and technological properties. These mathematical models are sophisticated in solutions and there are defined problems in solving the structure of complex automated lines by the criterion of the productivity rate alone. The scheme of the section-based automated line with a parallel–serial structure that is designed with q serial and p parallel workstations segmented into n serial sections with embedded buffer B is represented in Figure 5.13.

The properties of such lines are represented by combined properties of the automated line of parallel structure and the automated lines of serial structure segmented on sections with buffers that are presented in Chapters 3 and 4. The failure of any workstation in such structured automated line leads to a downtime of one section in the entire line, while other sections go on working. This automated line with parallel–serial structure

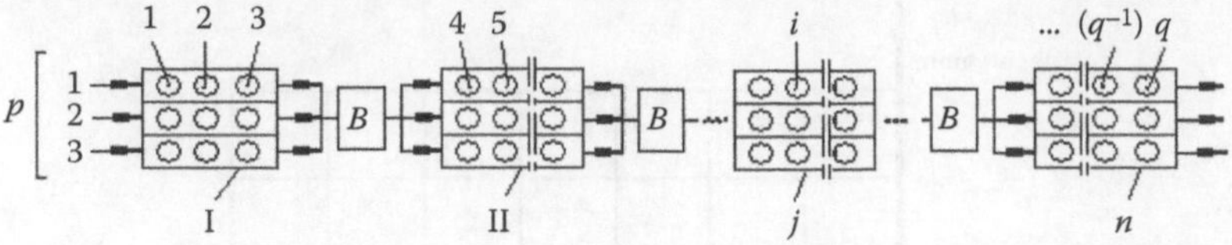

Figure 5.13 An automated line of the parallel–serial structure with p parallel lines of q serial workstations and n sections with buffers B.

possesses properties that depend on the number automated lines of parallel and serial structure, and sections with buffers. Combining serial and parallel workstations leads to a growth in the downtimes of the section as product of the number of parallel and serial workstations, i.e. $pq\lambda_s$ that reduces the productivity rate of the automated line of complex structure. The solution that automated line is segmented on the sections with buffers leads to reduction in the downtimes of sections, i.e. $pq\lambda_s/n$.

The actual productivity rate for the automated lines of parallel–serial structure segmented on sections with buffers possess some specific properties and limitations due to technical and technological reasons. These properties combine the properties of the automated line serial–parallel structure and the automated lines of serial structure segmented on sections with buffers. The maximal productivity rate with the increase in the number of parallel workstations leads to a decrease in the number of serial workstations. However, this dependency is different, because the structured automated line segmented on sections with buffers possess own properties that changes the diagram of the productivity rate, but tendency is the same. The equation for the productivity rate of a section-based automated line of parallel–serial structure segmented on sections with buffers depends on the structure of the line, the number of serial q, parallel p workstations, the number of sections n and the capacity of the buffers. Mathematical model of productivity rate for such structured line with buffers of limited capacity is a result of combining Eqs. (3.2), (4.5) and (4.17) and expressed by the following expression:

$$Q = \frac{p}{\dfrac{t_{mo}}{q} f_c + t_a} \times \frac{1}{1 + m_r \left[\dfrac{pf_s \displaystyle\sum_{i=1}^{q} \lambda_{s.i}}{n} + \Delta\lambda_{f(i-k)} + \cdots + \Delta\lambda_{f(i-2)} + \Delta\lambda_{f(i-1)} + \Delta\lambda_{b(i+1)} + \Delta\lambda_{b(i+2)} + \cdots + \Delta\lambda_{b(i+n)} + \lambda_{bf} + \lambda_c + \lambda_{tr} \right]}$$

$$(5.13)$$

For the automated line of serial–parallel structure and sections with big capacity, the equation for productivity rate is represented by the following simplified expressions of Eq. (5.12):

$$Q = \frac{p}{\frac{t_{mo}}{q} f_c + t_a} \times \frac{1}{1 + m_r \left[\frac{pf_s \sum\limits_{i=1}^{q} \lambda_{s.i}}{n} + \lambda_{bf} + \lambda_c + \lambda_{tr} \right]} \tag{5.14}$$

where all parameters are as specified earlier.

Equations (5.13) and (5.14) make it possible to decide the mathematical task of optimisation of the serial–parallel structure for the automated line, which means finding the optimal number of serial workstations q, the number of parallel workstations p and the number of sections n by criterion of maximum productivity. This task is solved by numerical solutions of Eqs. (5.13) and (5.14) with multi-variable parameters by method of phased approach. Since there are three variables q, p and n, it is necessary to compute Eqs. (5.13) and (5.14) with respect to one parameter q, p or n when other parameters are constant. This mathematical approach can give an answer to questions of whether or not there is an optimal number of structural components for the automated line of serial–parallel structure segmented on sections with buffers. Equations (5.13) and (5.14) for the structured automated lines with buffers of limited and big capacity are similar, but the first one is cumbersome in computing. For the following analyses, computing the productivity rate is used for the automated line of serial–parallel structure segmented on sections with buffers of big capacity.

A working example

The automated line of parallel–serial structure segmented on sections with buffers of big capacity is designed with the technical and timing data referred to one product and is shown in Table 5.7. The change in productivity rate versus the change in the number of serial workstations and sections should be computed. Results represent the diagram of productivity rate for the automated line with a complex structure.

Substituting data presented in Table 5.7 into Eq. (5.12), computing and transformation yield the result of the automated line productivity as functions of the number of serial stations q, the number of parallel workstation p and the number of parallel–serial sections n in the line.

$$Q = \frac{p}{\frac{30}{q} \times 1.1 + 0.3} \times \frac{1}{1 + 3 \times \left[\frac{p_s \times 1.2 \times 0.03 \times q}{n} + 7.0 \times 10^{-12} + 5.0 \times 10^{-8} + 6.0 \times 10^{-8} \right]}$$

The phased computing of the productivity rate with the change in the number of workstations q is implemented for the constant number of sections n for three structures of the automated lines. The changes in the productivity rate are represented graphically in Figures 5.14 and 5.15.

Table 5.7 Technical data of the automated line of serial–parallel structure segmented on sections with buffers of big capacity

Title	Data
Total machining time, t_{mo} (min)	30
Auxiliary time, t_a (min)	0.3
Correction factor for bottleneck workstation, f_c	1.1
Correction factor for bottleneck section, f_s	1.2
Average failure rate of the workstation, λ_{sav} (per min)	0.03
Failure rate of the control system, λ_c (per min)	5.0×10^{-8}
Failure rate of the transport system, λ_{tr} (per min)	6.0×10^{-10}
Failure rate of the buffer, λ_b (per min)	7.0×10^{-12}
Mean repair time, m_r (min)	3.0
Number of parallel workstations, p_s	2; 3; 4
Number of serial workstations, q	4,…,60
Number of sections, n	2; 3; 4

Figure 5.14 demonstrates the curves of the productivity change in automated lines of parallel–serial structure with two parallel workstations p, segmented into n sections with increase in the number of workstations q. The maximal productivity rate for the automated lines becomes displaced to the right with increase in the numbers of sections n and workstations q. This result confirms the tendency of changes in the productivities of the automated lines of serial structure and segmented on sections.

Figure 5.15 demonstrates the curves of the productivity change in automated lines of parallel–serial structure with p parallel workstations, segmented into n sections with increase in the number of workstations q. The maximal productivity rate for the automated lines

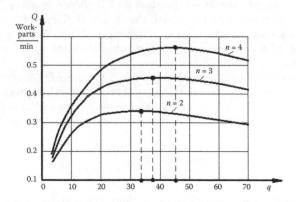

Figure 5.14 Productivity change in the automated line of a parallel–serial structure with buffered sections n and two parallel workstations versus the number of serial workstations q.

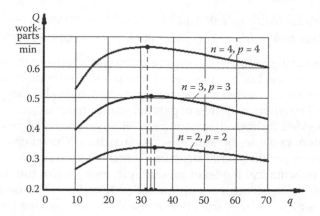

Figure 5.15 Productivity change in the automated line of a parallel–serial structure with buffered sections n and p parallel workstations versus the number of serial workstations q.

becomes displaced slightly to the left with increase in the numbers of sections n and workstations q. This result confirms the tendency of changes in the productivities of the automated lines of serial–parallel structure and segmented on sections.

The diagrams in Figures 5.14 and 5.15 show that increase in the number of parallel workstations p and sections n leads to increase in the productivity rate of a line. In addition, when increasing the number of serial stations q, the productivity rate is variable, leading to the optimal number of serial stations q that gives maximal productivity rate for the given number of parallel workstations p and sections n. However, when increasing the parallel workstations p and number of sections n, the number of optimal serial workstations q is decreased slightly.

Equations (5.13) and (5.14) give mathematical solutions for the optimal number of workstations q that change with the change of the number of sections n. The analytical results of the number of serial workstations q cannot be an integer number for a different number of parallel workstations p and sections n. From these results, it is necessary to take the nearest integer number of serial workstations q in the automated line. The maximal productivity rates for the considered examples demonstrated in Figures 5.14 and 5.15 by the spots that yield an optimal number of serial workstations.

Simplified equations of the productivity rates for automated lines of parallel–serial structure segmented on sections with equal failure rates for serial workstations based on Eqs. (4.18), (5.10) and (5.11) enable defining analytically the optimal structures for considered constructions of the automated lines. However, the results of computing can give big deviation from results obtained by accurate mathematical models.

5.2.2 Productivity rate of an automated line of parallel–serial structure with buffers after each line of parallel structure

In manufacturing industries, there are different designs of the automated lines that combined in complex serial and parallel structures. Such automated lines are connected by transport means, which also can play a role of buffers. The automated lines of parallel structure arranged in serial line with embedded buffers after each parallel automated line represent the typical automated lines of parallel–serial structure. The schematic picture of such line of parallel–serial structure is presented in Figure 5.16.

The mathematical model of productivity rate for the automated lines of parallel–serial structure with embedded buffers after each parallel structure is derived from analytical consideration presented for the automated line of serial structure segmented on sections (Section 4.4) and for the automated line of parallel structure (Section 3.1) and by Eqs. (5.13) and (5.14). Hence, the automated line of parallel–serial structure with buffers after each parallel line should be considered as a line segmented on sections where the number of automated lines of parallel structure q is equal to the number of sections n. Replacing components n by q in Eqs. (5.13), (5.14) and transformation give the following equations:

$$Q = \frac{p}{\frac{t_{mo}}{q} f_c + t_a} \times \frac{1}{\left[1 + m_r \left(\frac{pf_s \sum_{i=1}^{q} \lambda_{s.i}}{q} + \Delta\lambda_{f(i-k)} + \cdots + \Delta\lambda_{f(i-2)} + \Delta\lambda_{f(i-1)} + \Delta\lambda_{b(i+1)} + \Delta\lambda_{b(i+2)} + \cdots + \Delta\lambda_{b(i+n)} + \lambda_{bf} + \lambda_c + \lambda_{tr} \right) \right]}$$

$$(5.15)$$

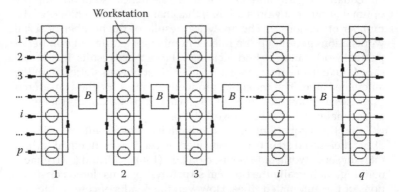

Figure 5.16 Scheme of automated production line of parallel–serial structure with embedded buffers *B* after each *q* automated line of parallel structure.

$$Q = \frac{p}{\dfrac{t_{mo}}{q} f_c + t_a} \times \frac{1}{1 + m_r \left[\dfrac{pf_s \sum\limits_{i=1}^{q} \lambda_{s.i}}{q} + \lambda_{bf} + \lambda_c + \lambda_{tr} \right]} \qquad (5.16)$$

where all parameters are as specified earlier.

The new Eqs. (5.15) and (5.16) for the productivity rate of an automated line of parallel structure arranged in serial construction with buffers after each line of parallel structure gives a solution for the maximal productivity rate. Productivity rate of this construction of automated line depends on the number of serial q and parallel p workstations, reliability of workstations and other units and the capacity of buffers. Analyses of Eqs. (5.15) and (5.16), and defined solutions for the productivity rate of the automated lines segmented on sections with buffers (Eqs. 5.10 and 5.11) enable stating the following. The change in the productivity rate for the automated line of parallel–serial structure with lines of parallel construction and buffers do not have optimal number of serial workstations that give maximal productivity rate. The change in productivity rate of considered construction of the automated line is similar to the serial line with buffers after each workstation, i.e. it comes asymptotically to the maximal value.

5.3   Productivity rate of rotor-type automated lines

Industrial processes in different areas such as pressworks (stamping, drawing, punching, trimming, coining, painting, etc.), assembling, liquid filling for different types of container (bottles, cans) that can be presented as surface and volume-type operations. These technological processes involve very short operation cycle time that are conducted in the rotor-type automatic machines. The single rotor-type automatic machine is a multi-station's machine of parallel structure. Generally, the rotor-type automated line belongs to manufacturing systems of serial–parallel structure (Figure 5.17). The main peculiarity of the rotor-type automated lines is that the technological process of the work parts at each workstation is executed with displacement at time. The rotor-type automated line has only one permanent mechanism for loading/unloading of the work parts in turn for all movable workstations. Generally, the application of rotor-type automated line is especially effective for machining of work parts with short operating cycles, and with high frequency of loading and unloading of work parts. Figure 5.17 represents the typical rotor-type automated line segmented on sections, and Figure 5.18 demonstrates the transport systems (buffers) between rotary machines.

Section 3.4.1 describes in detail and represents the mathematical model for productivity rate for the rotor-type automatic machine, which

Figure 5.17 The typical rotor-type automated lines.

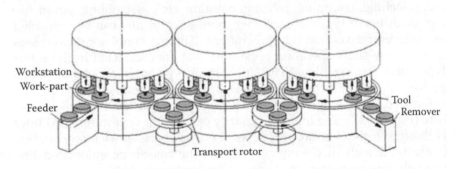

Figure 5.18 Transport systems that implement functions of the buffers between rotor-type automatic machines.

equation is used for the rotor-type automated line. The rotor-type automated line is designed for the complex technological processes for processing of operations that deal with surfaces and volumes. These complex processes are segmented on short elements of technological operations

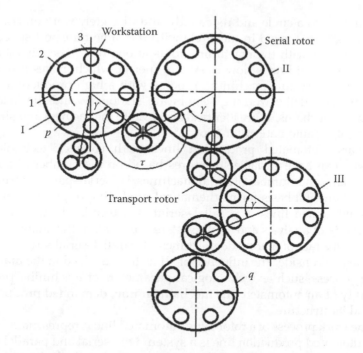

Figure 5.19 The scheme of the rotor-type automated line of parallel–serial structure with *p* parallel and *q* serial stations.

and executed according to the consequence of their fulfilment on rotor-type automated line. The demand to increase output of such an automated line leads to the creation of rotor-type automated lines embedded in one system. Figure 5.19 represents a sketch of rotor-type automated line that enable easy tracing of its work.

At the periphery of the working rotor, it has a continuous transport motion, disposed working units or workstations equipped with tool blocks for fulfilling the assigned operations. During the rotation of working rotor by means of the transport rotor, the work parts are loading into the working area of workstations. At sector $2\pi - \gamma$, the tool approaches the work parts rapidly, performs technological displacement (stamping, drawing, assembling etc.) and the tool is withdrawn. At the time of these motions, the work part is produced. At sector γ, the tool block is moved back in the open and initial position, the tool inspected, cleaned or replaced, and a new work part loaded. Sector $2\pi - \gamma$ fulfils the machining process (t_m) and auxiliary motions (t_a).

The rotor-type automated line is characterised by the fact that the workstations and tools for machining of work parts are moved continuously in a circle and repeatedly and discretely work on the work parts, which are moved in the circle with the same transport speed and having contact with the machining tools at one time. The advantage of rotary automated line is slow in system inertia, which makes it possible to conduct short-time machining process with a high productivity rate. Furthermore, all the working sectors are stationary, and the loading/ unloading mechanisms are identical and lead to the easiness and simplicity in creating automatic rotor-type lines.

Rotary automated production lines with complex technological processes can have variable structures based on the number of parallel workstations and number of serial machines, i.e. serial–parallel structure. The fundamental basis for mathematical modelling of productivity rate for all automated lines of parallel–serial structure is identical, because all methods of analysis and synthesis are identical. In the analysis of the productivity rate of automatic machines of parallel–serial structure, it is necessary to consider the influence of all factors involved in the manufacturing process, such as technological parameters of machining process, reliability of an automated line mechanism unit, demanded productivity rate and its structure.

The work process of a rotor-type automated line is represented earlier. This automated production line is a system with serial and parallel workstations arranged according to a certain structure, which depends on a technological process of machining work parts. Any failure, in either serial or parallel workstations, leads to stoppage of an entire rotary automated line, due to the mechanical hard joining of all mechanisms.

Practically, for industrial machines with complex designs, it is assumed that the reliability of each parallel machine tool or workstation in the system is the same. In the case of the rotor-type automated line, one principle is accepted for serial and parallel workstations: the system fails if at least one of the serial or parallel workstations fail. The failure of common mechanisms of machines with complex designs is considered separately. The result of this principle is that the increase in time losses according to Eq. (5.7) is expressed as follows:

$$\frac{p_s\left(\sum_{i=1}^{q}\theta_{s.i}+\theta_t\right)}{z}+\frac{\theta_{cs}}{z}=p_s\frac{\sum_{i=1}^{q}\theta_{s.i}+\theta_t}{z}+\frac{\theta_{cs}}{z}=m_r\left[p_s\left(\sum_{i=1}^{q}\lambda_{s.i}+\lambda_{tr}\right)+\lambda_{cs}\right]$$

$$(5.17)$$

where θ_s is the idle time of one station, θ_t is the idle time of one transport rotor; θ_{cs} is the idle time of common mechanisms like control system; p_s is the number of parallel workstations; q is the number of serial workstations and other parameters are as specified earlier.

The $\lambda_{s.i,}$ λ_{tr} and λ_{cs} are failure rate of workstation i, transport rotor and common mechanisms, respectively. The number of workstations in the transport rotor is less than the number of parallel workstations in the working rotor, but the intensiveness of the operation in workstations of one transport rotor is higher. Hence, the number of parallel stations p_s multiplies the time losses due to the reliability of one transport rotor.

To derive the equation of productivity rate for the rotary auto-mated line, it is important to reconsider calculation of cycle time and to include time duration for transportation of work parts to the next rotor machine according to the technological process [14]. The angle γ in the rotary machine is constructive angle that is necessary to compile the transport rotor with the machining one. Designers tend to make angle γ as small as possible, because this is not a productive angle and considered as constructive necessity for assembling of rotary units and maintaining them. To produce analytical dependency of productivity rate of automated line, it is necessary to consider other components. The machining process of a work part in one rotor-type machine is implemented over angles $(2\pi - \gamma)$ and the process is fulfilled during cycle time, T.

Design features of the rotor-type automated lines include transport rotors, which transfer work parts from one rotor machine to the next. For transportation of machined work parts, the time spent is t_{tr}, which is not less, and frequently, is bigger than the auxiliary time t_a of a single worksta-tion of the rotor-type machine and should be considered in the expression for productivity rate. Obviously, the productivity rate of the rotor-type automated line with parallel–serial structure should include also the time t_{tr} for inter-rotary machines transportations of work parts into the lines having q serial rotary machines.

Transportation time of a work part by the transport rotor is calculated as follows: $t_{tr} = p_\tau l/V$ where l is constructive length of circular arc between two rotor neighbouring parallel workstations; $V = l(p_s - p_\gamma)/T$ is the tan-gential velocity of the circular motion for parallel workstations, defined earlier (Section 3.4.1) and p_τ is number of workstations located on angle τ of transport rotor (Figure 5.3). The angle τ is similar as the angle γ, i.e. these angles are out of the productive angle $(2\pi - \gamma)$ on the working rotor. Substituting these expressions and transformation give the equation for the transportation time, thus

$$t_{tr} = \frac{Tp_\tau}{p_{(\alpha+\beta)}} = (t_m + t_a)\frac{p_\tau}{p_{(\alpha+\beta)}} = (t_m + t_a)\frac{p_\tau}{p_s - p_\gamma} \qquad (5.18)$$

Incorporating Eq. (5.18) of transportation time into Eq. (3.9), transformation and simplification the equation for productivity of the rotor-type automatic machine with transport rotor is given by the next equation:

$$Q = \frac{p}{(t_m + t_a)\left(2 + \dfrac{p_\gamma - 1}{p_s - p_\gamma}\right) + t_{tr}} \times \frac{1}{1 + m_r\left(p_s[\lambda_s + \lambda_{tr}] + \lambda_{cs}\right)}$$

$$= \frac{p}{(t_m + t_a)\left(2 + \dfrac{p_\gamma + p_\tau - 1}{p_s - p_\gamma}\right)} \times \frac{1}{1 + m_r\left(p_s[\lambda_s + \lambda_{tr}] + \lambda_{cs}\right)} \tag{5.19}$$

where p_τ is number of workstations located on angle τ of transport rotor (Figure 5.3) and $p = p_s$ is number of work parts machined in the rotor machine. Other parameters are as specified earlier.

Analysis of the rotor-type automatic machines and designs that embedded into the automated line demonstrates the numbers of parallel workstations p_γ and number of parallel workstations in the transport rotors, where p_τ is not a variable component. It is integer numbers in rotor-type automatic machine from construction point of view and located on angle γ, and τ accordingly as mentioned earlier. The total number of operating parallel workstations p_s is always more than p_γ ($p_s > 2p_\gamma$) and cannot be less than one from the construction point of view. When the rotor machine in the automated line does not have idle parallel workstations p_γ, and does not have the transport rotor p_τ ($p_\gamma = 0$; $p_\tau = 0$), Eq. (5.19) is transforming to the following form:

$$Q = \frac{p}{(t_m + t_a)\left(2 - \dfrac{1}{p_s}\right)} \times \frac{1}{1 + m_r\left(p_s\lambda_m + \lambda_{cs}\right)} \tag{5.20}$$

where all parameters are as specified earlier.

The equation of the productivity rate for the rotor-type automatic line of parallel–serial structure combines the properties of the automated line of serial structure (Eq. 4.6) and the rotor-type automatic machine (Eq. 3.10). New equations of productivity rate for the rotor-type automated line is represented as follows:

$$Q_{pq} = \frac{p}{\left(\dfrac{t_{mo}}{q} f_c + t_a\right)\left(2 + \dfrac{p_\gamma + p_\tau - 1}{p_s - p_\gamma}\right)} \times \frac{1}{1 + m_r\left[p_s(\sum\limits_{i=1}^{q} \lambda_{s.i} + \lambda_{tr}) + \lambda_{cs}\right]} \tag{5.21}$$

where all parameters are as presented earlier.

Equation (5.21) for productivity rate of the rotor-type automated line has components that represent structure, which is the number of serial working rotors q and parallel workstations p_s (Figure 5.19). This equation includes parameters of the technological process t_{mo}, design t_a and components of failure rates due to technical reasons with the automated line that reflects its reliability and structure (λ_{cs}, λ_s, λ_{tr}).

Analysis of Eq. (5.21) demonstrates that one component, whose expression may be given by the form

$$\left(\frac{t_{mo}}{q}+t_a\right)\left(2+\frac{p_\gamma+p_\tau-1}{p_s-p_\gamma}\right)=T\left(2+\frac{p_\gamma+p_\tau-1}{p_s-p_\gamma}\right)=Td \text{ is the processing time}$$

of p work parts, where $d=2+\dfrac{p_\gamma+p_\tau-1}{p_s-p_\gamma}$ is the cycle time displacement factor.

To prove that Eq. (5.21) is correct, it can be represented in the form of symbols of time losses by the following equation:

$$Q=\frac{p}{\left(\frac{t_{mo}}{q}f_c+t_a\right)\left(2+\frac{p_\gamma+p_\tau-1}{p_s-p_\gamma}\right)}\times\left(\frac{1}{1+\frac{qp_s(t_s+t_{tr})+t_{cs}}{Td}}\right)=Q_cA, \quad (5.22)$$

where the first component of Eq. (5.22) is the displaced productivity rate per cycle of the rotor-type automated line:

$$Q_c=\frac{p}{\left(\frac{t_{mo}}{q}+t_a\right)\left(2+\frac{p_\gamma+p_\tau-1}{p_s-p_\gamma}\right)} \quad (5.23)$$

The second component of Eq. (5.22) is the availability of the rotor-type automated line:

$$A=\frac{1}{1+\dfrac{qp_s(t_s+t_{tr})+t_{cs}}{Td}}$$

$$=\frac{1}{1+\left[\left\{qp_s(t_s+t_{tr})+t_{cs}\right\}\Big/T\left(2+\frac{p_\gamma+p_\tau-1}{p_s-p_\gamma}\right)\right]} \quad (5.24)$$

where all parameters are as specified earlier.

The availability A of an automated line includes the time losses (t_{cs}, t_s, t_{tr}) due to technical reasons, which reflect the reliability parameter of the mechanisms and workstations, and structural parameters of the line. However, the displacement of the cycle time (d) in the machining process at each workstation does not reflect the availability level of the completely automated system. This statement is proven as follows.

The time losses due to technical reasons (t_{cs}, t_s, t_{tr}) is expressed by the standard attributes of the reliability of a machine. The time losses of the rotor-type automated line due to technical reasons are represented by Eq. (5.17):

$$qp_s(t_s + t_{tr}) + t_{cs} = qp_s\left(\frac{\theta_s}{z} + \frac{\theta_{tr}}{z}\right) + \frac{\theta_{cs}}{z} \tag{5.25}$$

where all parameters are as specified earlier.

The idle times of the mechanisms and machines of an automated line due to technical reasons are expressed by attributes of reliability, $\theta_i = m_r b$, where m_r is the mean time to repair and b is the number of failures. The machine's work time is expressed by the equation: $\theta_w = m_w b = b/\lambda$, where λ is the machine failure rate. The number of products produced is expressed as $z = \theta_w/Td$. Transforming the expression of idle times can be done similar to that shown earlier for a single machine, Eq. (2.2).

The time losses of any i mechanism in an automated line is represented by standard attributes of reliability via modified Eq. (2.4):

$$t_{t.i} = \frac{\theta_i}{z} = \frac{m_r k}{\theta_w / T_t} = \frac{m_r k}{m_w b / T} = Tm_r \lambda \frac{k}{b} = Tm_r \lambda_i$$

where $t_{t.i}$ is the time losses of any i mechanism in an automated line; $\theta_i = m_r k$ is idle time due to the considered mechanism; k is number of failures in considered i mechanism; $\theta_w = m_w b$ is the work time of an automated line, where b is the number of failures in an automated line; T is cycle time; $m_w = 1/\lambda$ is work mean time of an automated line; λ is failure rate of an automated line; $b = \Sigma k_i$ is the number of failures in the automated line equal to the sum of the failures of mechanisms and workstations in the line; $\lambda_i = \lambda k/b$ is the failure rate of considered i mechanism; other parameters are as specified earlier.

In the industry, the attribute of mean time to repair m_r random failures of machine units is accepted as constant with values of several minutes. However, the attribute of machine failure rate is different for different machine units. Hence, the equation of the rotor-type automated line's availability should include λ_{cs}, λ_s and λ_{tr}, that is, the failure rates for control mechanisms, rotor workstations and transport rotor, respectively.

Substituting the reliability attributes and other parameters represented earlier into Eq. (5.22) and transformation gives the equation for the availability of the rotor-type automated line as follows:

$$A = \cfrac{1}{1 + \left[\left\{ p_s q \left(\cfrac{\theta_s}{\theta_w / (Td)} + \cfrac{\theta_{tr}}{\theta_w / (Td)} \right) + \cfrac{\theta_{cs}}{\theta_w / (Td)} \right\} \Big/ Td \right]}$$

$$= \cfrac{1}{1 + \left[\left\{ p_s q \left(\cfrac{m_r b_s \lambda_s Td}{b_s} + \cfrac{m_r b_{tr} \lambda_{tr} Td}{b_{tr}} \right) + \cfrac{m_r b_{cs} \lambda_{cs} Td}{b_{cs}} \right\} \Big/ Td \right]} \qquad (5.26)$$

$$= \cfrac{1}{1 + m_r \left[p_s q (\lambda_s + \lambda_{tr}) + \lambda_{cs} \right]}$$

where all parameters are as specified earlier.

Equation (5.26) contains standard attributes for the reliability of mechanisms (m_r, λ_i), and parameters of structure and design for the rotor-type automated line (p_s, q). Equation (5.26) demonstrates that availability depends on the standard attributes of reliability and on the structural parameter of a rotor-type automated line. Substituting Eq. (5.26) into Eq. (5.20) with corrections on differences of workstation's failure rates and the machining time (Eq. 4.3), the productivity rate of the rotor-type automated line will have the equation that contains all the technical parameters and availability of the line. The availability of the rotor-type automated line includes the mean time of repair (m_r), machine failure rates (λ_i) and its structure (p_s, q):

$$Q_{pq} = \cfrac{p}{\left(\cfrac{t_{mo}}{q} f_c + t_a \right)\left(2 + \cfrac{p_\gamma + p_t - 1}{p_s - p_\gamma} \right)} \times \cfrac{1}{1 + m_r \left[p_s \left(\sum_{i=1}^{q} \lambda_{s.i} + \lambda_t \right) + \lambda_{cs} \right]}$$

$$(5.27)$$

where p is the number of work parts (each station carries one work part where the number of work part is equal to the number of workstations, $p = p_s$; q is the number of serial workstations (working rotor-type machine); t_{mo} is the machining time of the total technological process; λ_s, λ_{tr} and λ_{cs} are the failure rate due to the reliability of the workstation's mechanisms, transport rotor and control system, respectively; p_τ is the number of workstations located at the transport sector τ of the transport rotor; p_γ is the number of workstations located at the sector γ of the working rotor (Figure 5.16). Other parameters are as presented earlier.

Equation for productivity rate (5.27) of the rotor-type automated line has components that represent structure, which is the number of serial working rotors (q) and parallel workstations (p_s). This equation includes parameters of the technological process (t_{mo}), design (t_a) and components of reliability due to technical reasons that reflect its structure of the rotor-type automated line (λ_s, λ_{tr}, λ_{cs}). Equation (5.23) for the productivity rate per cycle of the rotor-type automated line and Eq. (5.26) for its availability make it possible to solve important engineering problems. These equations allow for calculating the required attributes of reliability in the rotor-type automated line according to the requested productivity rate or vice versa. If the values of reliability indices in the automated line components are known, it is easy to calculate the productivity rate.

The rotor-type automated line is designed by combining with serial rotor-type machines q according to the technological process of machining the work parts. Each serial rotor-type machine can have different machining and cycle times. The problem of balancing of technological process for serial workstations at rotor-type automated line does not create difficulties. The differences in cycle times of serial workstations are compensated by increasing or decreasing the number of parallel workstations in the rotor-type machines. However, the productivity rate per cycle on each rotor-type machine of the automated line should be the same. This statement is represented by the following equation:

$$Q_c = \frac{p_1}{T_1\left(2+\dfrac{p_\gamma+p_\tau-1}{p_{s.1}-p_\gamma}\right)} = \frac{p_2}{T_2\left(2+\dfrac{p_\gamma+p_\tau-1}{p_{s.2}-p_\gamma}\right)} = \frac{p_3}{T_3\left(2+\dfrac{p_\gamma+p_\tau-1}{p_{s.3}-p_\gamma}\right)} = \cdots$$

$$= \frac{p_i}{T_i\left(2+\dfrac{p_\gamma+p_\tau-1}{p_{s.i}-p_\gamma}\right)} = \cdots = \frac{p_q}{T_q\left(2+\dfrac{p_\gamma+p_\tau-1}{p_{s.q}-p_\gamma}\right)} \tag{5.28}$$

where p_i is the number of the parallel workstations and T_i is the cycle time at the rotor-type machine i, $[2+(p_\gamma+p_\tau-1)/(p_{s.i}-p_\gamma)]$ is the cycle time displacement factor, the number of workstations located at angles γ and τ is accepted as constant; other parameters are as specified earlier.

Practically, the rotor-type of automatic machines with different number of parallel workstations on each rotor-type machine have different reliabilities. The failure of common mechanisms of machines with complex designs is considered separately. The productivity rate of the rotor-type automated line should be considered by the rotor-type machine that has maximum number of parallel workstations, i.e. can have high failure rate and hence it is bottleneck rotor-type serial machine. However, it is not regular practice; the rotor-type machine with high failure rate can be the machine with less parallel workstations. This rotor-type machine will have high frequency of rotation and small number of workstations that can lead to high failure rate. Hence, any rotor-type machine with different number of parallel workstations can be a bottleneck machine in the automated line. Which rotor-type machine is bottleneck depends on its reliability attributes.

Equation (5.27) for productivity rate based on the different failure rates of the rotor-type automated lines workstations is effectively used for the optimisation analysis of the rotor-type automated line parameters. Analysis of this equation demonstrates that it has the extreme of this function. It means that it is possible to find the values of the two variables p_s and q that give the maximal value of functions. Numerical solution of Eq. (5.27) can demonstrate optimal number of serial and parallel workstations that yields the maximal productivity rate. For simplification of analysis, it can be used for analytical solutions. In such case, it is accepted that the equal reliability of serial workstations for Eq. (5.27) will have the following expression:

$$Q_{pq} = \frac{p}{\left(\dfrac{t_{mo}}{q}f_c + t_a\right)\left(2 + \dfrac{p_\gamma + p_t - 1}{p_s - p_\gamma}\right)} \times \frac{1}{1 + m_r\left[qp_s(\lambda_s + \lambda_t) + \lambda_{cs}\right]} \quad (5.29)$$

Then, the mathematical task of optimisation of the structure of rotor-type automated lines by criterion of maximal productivity is decided. In this study, the partial derivative for Eq. (5.29) is used since only one optimal point is expected for each independent variable. Maximum productivity rate is reached by solving the first partial derivatives of Eq. (5.29) with two variables q and p.

Differentiating this equation with first variable q yields

$$\frac{dQ}{dq} = \frac{p}{\left(\dfrac{t_{mo}}{q}f_c + t_a\right)\left(2 + \dfrac{p_\gamma + p_t - 1}{p_s - p_\gamma}\right)} \times \frac{1}{1 + m_r[qp_s(\lambda_s + \lambda_{tr}) + \lambda_{cs}]}$$

giving rise to the following

$$q_{opt} = \sqrt{\frac{t_{mo}f_c([1/m_r] + \lambda_{cs})}{p_s(\lambda_s + \lambda_{tr})t_a}} \quad (5.30)$$

where all parameters are as specified earlier.

Equation (5.30) represents the optimal number of serial workstations, which gives maximal productivity rate for the rotor-type automated line with the given number of parallel workstations. Optimal number of serial workstation in the rotary automated line depends on the number of parallel workstations, failure rates of workstation, transport and control systems and bottleneck machining and auxiliary times. Substituting

Eq. (5.30) into Eq. (5.29) and transformation yield the following equation for the maximal productivity rate of the rotor-type automated line:

$$Q_{max} = \frac{1}{\left(\sqrt{\dfrac{(\lambda_s + \lambda_{tr})t_{mo}f_c t_a}{p_s t_{mo} f_c([1/m_r] + \lambda_{cs})}} + t_a \right)\left(2 + \dfrac{p_\gamma + p_\tau - 1}{p_s - p_\gamma} \right)}$$

$$\times \frac{1}{1 + m_r \left[q \sqrt{\dfrac{t_{mo}f_c([1/m_r] + \lambda_{cs})(\lambda_s + \lambda_{tr})}{p_s t_a}} + \lambda_{cs} \right]}$$

(5.31)

As far as the rotor-type automated line belongs to the lines of parallel–serial structures, i.e. these constructions of the line do not have optimal number of parallel workstations confirmed in Section 5.2.

A working example 1

The rotor-type automated line with technical and technological data represented in Table 5.8 can be designed in variants. Calculate the productivity rates for the given data of the rotor-type automated line. The results are presented in diagrams as the change in the productivity rate versus the change in the number of parallel p_s and serial q workstations.

Equations (5.29) and (5.30) calculate the productivity rates with variable parameters in the rotor-type automated line, which are the number of serial and parallel workstations.

Substituting the initial data of Table 5.8 into Eqs. (5.30) and (5.29), and the following calculated results are represented in Figure 5.20. The optimal number of serial workstations q_{opt} that gives the maximal productivity rate Q_{max} is as follows:

- For the automated line of six parallel workstations

$$q_{opt} = \sqrt{\frac{t_{mo}f_c([1/m_r] + \lambda_{cs})}{p_s(\lambda_s + \lambda_{tr})t_a}}$$

$$= \sqrt{\frac{1.0 \times 1.2 \times ([1/2.0] + 0.0005)}{6 \times (0.009 + 0.0045) \times 0.1}} = 8.61$$

$$Q_{max} = \frac{p}{\dfrac{t_{mo}}{q_{opt}}f_c + t_a} \times \frac{1}{1 + m_r(p_s q_{opt}\lambda_{sav} + \lambda_{tr} + \lambda_{cs})}$$

$$= \frac{6}{\dfrac{1.0}{8.61} \times 1.2 + 0.1} \times \frac{1}{1 + 2.0(6 \times 8.61 \times 0.009 + 0.0045 + 0.0005)}$$

$$= 12.908 \text{ work part/min}$$

Table 5.8 Technical data of the rotor-type automated line

Title	Data
Total machining time, t_{mo} (min)	1.0
Auxiliary time, t_a (min)	0.1
Correction factor for bottleneck workstation, f_c	1.2
Average failure rate of workstation, s_{av} (per min)	0.009
Failure rate of the control system, λ_{cs} (per min)	0.0005
Failure rate of the transport system, λ_{ts} (per min)	0.0045
Mean repair time, m_r (min)	2.0
Number of parallel workstations, p_s	6,...,10
Number of workstations in the idle angle γ	2
Number of workstations in the transport angle τ	2
Number of serial workstations, q	6,...,14

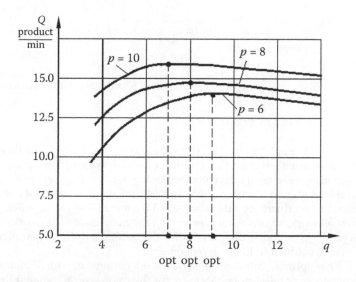

Figure 5.20 Productivity rate of the rotor-type automated line.

- For the automated line of eight parallel workstations

$$q_{opt} = \sqrt{\frac{t_{mo} f_c ([1/m_r] + \lambda_{cs})}{p(\lambda_s + \lambda_{tr})t_a}}$$

$$= \sqrt{\frac{1.0 \times 1.2 \times ([1/2.0] + 0.0005)}{8 \times (0.009 + 0.0045) \times 0.1}} = 7.457$$

$$Q_{max} = \frac{p}{\frac{t_{mo}}{q_{opt}}f_c + t_a} \times \frac{1}{1 + m_r(p_s q_{opt}\lambda_{sav} + \lambda_{tr} + \lambda_{cs})}$$

$$= \frac{8}{\frac{1.0}{7.45} \times 1.2 + 0.1} \times \frac{1}{1 + 2.0(8 \times 7.45 \times 0.009 + 0.0045 + 0.0005)}$$

$$= 14.7 \text{ work part}/\text{min}$$

- For the automated line of 10 parallel workstations

$$q_{opt} = \sqrt{\frac{t_{mo}f_c([1/m_r] + \lambda_{cs})}{p(\lambda_s + \lambda_{tr})t_a}}$$

$$= \sqrt{\frac{1.0 \times 1.2 \times ([1/2.0] + 0.0005)}{10 \times (0.009 + 0.0045) \times 0.1}} = 6.67$$

$$Q_{max} = \frac{p}{\frac{t_{mo}}{q_{opt}}f_c + t_a} \times \frac{1}{1 + m_r(p_s q_{opt}\lambda_{sav} + \lambda_{tr} + \lambda_{cs})}$$

$$= \frac{10}{\frac{1.0}{6.67} \times 1.2 + 0.1} \times \frac{1}{1 + 2.0(10 \times 6.67 \times 0.009 + 0.0045 + 0.0005)}$$

$$= 16.14 \text{ work part}/\text{min}$$

Figure 5.20 depicted by Eq. (5.29) demonstrates that productivity rate grows with increase in number of serial workstations q at the beginning, reaches its maximal value for defined number of serial workstations and then decreases. In addition, there is a tendency that maximal productivity rate is achieved by increasing the number of parallel workstations p_s and decreasing the number of serial workstations q. This trend is the same as for the automated line of parallel–serial structure and linear arrangement.

The optimal number of serial workstations q with different numbers of parallel workstations p_s for the rotor-type automated line calculated by Eq. (5.30) gives the following results:

For $p_s = 6$, $q = 8.61$, after rounding $q = 9$;
For $p_s = 8$, $q = 7.45$, after rounding $q = 8$;
For $p_s = 10$, $q = 6.67$, after rounding $q = 7$;

Relevant equations are represented, respectively, for productivity rate of rotor-type automated line, optimal number of serial workstations that gives the maximum productivity rate. These equations enable the calculation the technical characteristics and productivity rate as a function of number of parallel and serial workstations. Derived equations enable evaluation on the variations of productivity rate of rotor-type automated line with different structures.

A working example 2

The rotor-type automated line is designed with $q = 5$ serial rotor machines and $p_s = 8,\ldots,16$ parallel stations in one rotor machine (Figure 5.16). We assume the cyclic productivity is given as $Q = 2.1$ product/min. The machining time t_m at each station, the number of idle stations $p\gamma$ in the rotor machine and the number of stations p_{tr} in the transport rotor are given in Table 5.9. The number of parallel stations in the rotor machines is calculated using Eq. (5.29). Substitution of the initial data of the first serial station (Table 5.9) into Eq. (5.29) and transformation give the following ($p = p_s$):

$$2.1 = \frac{p}{(t_{m.1} + 0.2)\left(2 + \dfrac{3+4-1}{p_s - 3}\right)} \text{ or } p = 4.2(t_{m.1} + 0.2) + 3.$$

For each serial rotor machine, the following parameters are calculated:

- The number of parallel stations, $p_i = 4.2(t_{m.i} + 0.2) + 3$ (rounded to the nearest integer number)
- The machining correction factors, $f_{m.i} = t_{m.i}/t_{av}$
- The productivity rate per cycle time, $Q_i = \dfrac{p_i}{(t_{m.i} + t_a)\left(2 + \dfrac{p_\gamma + p_\tau - 1}{p_s - p_\gamma}\right)}$

The calculated and given technical and technological data for the rotor-type automated line are presented in Tables 5.9 and 5.10.

We assume that the bottleneck serial station (due to reliability) is described by $q = 3$ with $p_s = 16$ parallel stations. Substituting the initial

Table 5.9 Technical data for a rotor-type automated line

Title	q_i	p_i	$t_{m.i}$ (min)	$f_{m.i}$	Q_i (product/min)
Serial station	1	8	0.98	0.548	2.11
	2	10	1.45	0.812	2.12
	3	16	2.80	1.567	2.16
	4	12	1.90	1.064	2.14
	5	12	1.80	1.008	2.25
Total machining time, t_{mo} (min)					8.93
Average machining time t_{av} (min)					1.786
Auxiliary time, t_a (min)					0.2
Number of idle stations in the rotor machine, p_γ					3
Number of transport stations, p_τ					4

Table 5.10 Reliability attributes for a rotor-type automated line

Title	q	$\lambda_{s,i}$ (k/min)
Failure rate of the serial rotor machine q	1	40.0×10^{-3}
	2	35.0×10^{-3}
	3	45.0×10^{-3}
	4	30.0×10^{-3}
	5	50.0×10^{-3}
	Average	40.0×10^{-3}
Failure rate of the control system, λ_c		8.0×10^{-5}
Failure rate of the transport system, λ_{tr}		4.0×10^{-5}
Mean repair time $m_r = 3.0$ min		

data (Tables 5.9 and 5.10) into Eqs. (5.30), (5.31) and performing the calculations give Eqs. (Q_1) and (Q_2).

Substituting the initial data (Tables 5.9 and 5.10) into Eq. (5.27) gives Eq. (Q_1) with variable parameters q and λ_s, which depend on the serial rotor machines. The correction factors f_m should be omitted, because the number of parallel workstations in the serial rotors as corrected as the difference in cycle times. The number of parallel stations $p_s = 16$ is constant for all serial rotor machines, which contain smaller numbers of parallel stations. This assumption is correct because the intensiveness of the work of the parallel stations in other serial rotor machines is higher, and the total number of working stations is equal to the number of stations at the bottleneck rotor machine.

$$Q_1 = \frac{p}{\left(\dfrac{t_{mo}}{q} + t_a\right)\left(2 + \dfrac{p_\gamma + p_\tau - 1}{p_s - p_\gamma}\right)} \times \frac{1}{1 + m_r\left[p_s \sum\limits_{i=1}^{q}(\lambda_{s,i} + \lambda_{tr}) + \lambda_c\right]}$$

$$= \frac{16}{\left(\dfrac{8.93}{q} + 0.2\right)\left(2 + \dfrac{3+4-1}{16-3}\right)} \times \frac{1}{1 + 3.0\left[16\sum\limits_{i=1}^{q}(\lambda_s + 4.0\times 10^{-5}) + 8.0\times 10^{-5}\right]}$$

$$Q_2 = \frac{p}{\left(\dfrac{t_{mo}}{q} + t_a\right)\left(2 + \dfrac{p_\gamma + p_\tau - 1}{p_s - p_\gamma}\right)} \times \frac{1}{1 + m_r[p_s q(\lambda_s + \lambda_{tr}) + \lambda_c]}$$

$$= \frac{16}{\left(\dfrac{8.93}{q} + 0.2\right)\left(2 + \dfrac{3+4-1}{16-3}\right)} \times \frac{1}{1 + 3[16q(40\times 10^{-3} + 4.0\times 10^{-5}) + 8.0\times 10^{-5}]}$$

Equations (Q_1) and (Q_2) enable the calculations (Table 5.11) and result in the diagram (Figure 5.21) of the change in the productivity rate for

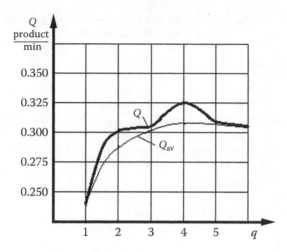

Figure 5.21 Productivity rate for a rotor-type automated line versus the number of serial and parallel workstations with different failure rates.

Table 5.11 Productivity rate for a rotor-type automated line with q serial and p parallel workstations

$p_s = 16; q$	1	2	3	4	5
Q (product/min)	0.243	0.302	0.302	0.325	0.307
Q_{av} (product/min)	0.243	0.287	0.301	0.307	0.307

the rotor-type automated line with a different number of serial q and parallel p stations and failure rates.

The maximum productivity rate $Q = 0.325$ product/min describes an automated line with $q = 4$ serial rotor machines (curve Q). The curve Q_{av} represents the change in productivity rate calculated by the average failure rates of the rotor machines. Equation (5.11) gives the optimal number of serial rotor machines, which determines the maximum productivity rate of the rotor-type automated line ($Q_{max} = 0.307$ product/min), which, together with average failure rate of the serial rotor machines, gives the following result:

$$q_{opt} = \sqrt{\frac{t_{mo} f_c [(1 / m_r) + \lambda_c + \lambda_{tr}]}{p_s t_a \lambda_s}}$$

$$= \sqrt{\frac{8.93[(1 / 3.0) + 8.0 \times 10^{-5} + 4.0 \times 10^{-5}]}{16 \times 0.3 \times 0.040}} = 4$$

There are differences in the results calculated by Eqs. (Q_1) and (Q_2). The simplified approach in the calculations yields a smaller maximum productivity rate than those produced from the detailed process of

Table 5.12 Technical data of the rotor-type and linear automated line of parallel–serial structure

Title	Data
Total machining time, t_{mo} (min)	1.0
Auxiliary time, t_a (min)	0.1
Correction factor for bottleneck workstation, f_c	1.2
Average failure rate of the workstation, λ_{sav} (per min)	$\lambda \times 10^{-4}$
Failure rate of the control system, λ_{cs} (per min)	$\lambda \times 10^{-8}$
Failure rate of the transport system, λ_{ts} (per min)	$\lambda \times 10^{-10}$
Mean repair time, m_r (min)	3.0
Number of serial workstations, q	4,…,12
Number of parallel workstations, p_s	10,…,20
Number of workstations in the idle angle γ	2
Number of transport workstations at the rotor angle τ	2
Availability, A	0.8,…,0.9

calculation [9,15]. This is the validation that simplified approach to the productivity rates of the automated lines does not give correct result.

A working example 3

The rotor-type and linear arrangement automated lines of parallel–serial structure.

The rotor-type (Figure 5.16) and linear (Figure 5.5) arrangement automated production lines can be designed with variants for which the technical data are presented in Table 5.12.

These normative data allow for the solving of engineering problems that concern the attributes of reliability, structural and technical parameters, and the productivity rate of the automated line. Substituting these data into Eqs. (5.29) and (5.7), the results of availability and productivity rates of the rotor-type and linear arrangements automated lines versus the failure rate are represented in Figure 5.22. The diagram shows that increase in the failure rate of the automated line decreases the magnitude of availability of the automated line. Increasing the number of parallel and serial workstations in the automated line reflects the magnitude of availability of the automated line: more workstations means availability with less value. Comparative analysis of the productivity rates demonstrates that the linear arrangement of the automated line gives bigger productivity rate than the rotor-type automated line.

If the availability accepted is $A = 0.85$, then

- An automated line with $q = 12$ serial and $p_s = 20$ parallel workstations should have a failure rate of $\lambda \leq 2.4 \times 10^{-4}$ failures/min.
- An automated line with $q = 8$ serial and $p_s = 15$ parallel workstations should have a failure rate of $\lambda \leq 5.0 \times 10^{-4}$ failures/min.
- An automated line with $q = 4$ serial and $p_s = 10$ parallel workstations has a high attribute of availability.

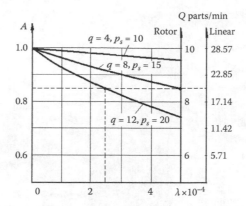

Figure 5.22 Productivity rate and availability of rotor-type and linear arrangement automated lines with q serial and p_s parallel workstations versus increase in the failure rate λ of the automated lines.

Two structural variants of the rotor-type and linear arrangement automated lines give the productivity rate $Q_t = 8.5$ work parts/min and 20.0 work parts/min, respectively. The automated line with $q = 4$ serial and $p_s = 10$ parallel workstations has quite a low failure rate and high reliability, and thus gives a higher productivity rate. In the manufacturing industry, the automated lines of parallel–serial structure and linear arrangements can have two and rarely three parallel automated lines; however, the machinery of the electronic industry can have 10–15 parallel automated lines.

5.3.1 Productivity rate of rotor-type automated lines segmented on sections with buffers

In industries, usually, the rotor-type automated lines are not designed with many serial workstations for the reason of complexity of dispositions for rotor-type machines in the automated line. Rotor-type automated lines with complex technological processes are systems of parallel–serial structures. The rotor machines present parallel workstations and the number of rotor machine presents serial workstations in sequence of implementation of the technological process for machining the work parts. Their structures can be variable and based on number of parallel workstations in the rotor machines and number of serial machines in the automated line. Most of the rotor-type automated lines are segmented on sections or designed with ability that each rotor-type machines work separately and relatively independent. Between the sections of rotor-type automated lines or rotor-type automatic machines are embedded transport mechanisms that play the role of buffers. Figure 5.23 represents the typical rotor-type automated line segmented on sections and Figure 5.24 the transport systems (buffers) between rotary machines.

Figure 5.23 The rotor-type automated line.

Figure 5.24 Transport systems are the buffers between rotor-type automatic machines.

A sketch of rotor-type automated line segmented on sections with buffers is represented in Figure 5.25a. A rotor-type automated line with buffers after each rotor machine is represented in Figure 5.25b.

The work process of a rotor-type automated line is shortly presented earlier. The automated line segmented on sections is a system with buffers, serial and parallel workstations arranged according to a certain structure that depends on a technological process of machining work parts. Any failure in the rotary machine or mechanisms of a section leads to a downtime of a section of rotary automated line only. Other sections work on filling and consuming of work parts from the buffers. The equation of the productivity rate for the rotor-type automated line of parallel–serial structure segmented on sections with buffers combines the properties of the rotor-type automated line (Eq. 5.31) and the automated line of parallel–serial structure segmented on sections with buffers of limited or big capacity (Eqs. 5.12 and 5.13). New equation of productivity rate for the rotor-type automated line with buffers of limited capacity is represented as follows:

$$Q = \frac{p}{\left(\dfrac{t_{mo}}{q} f_c + t_a \right)\left(2 + \dfrac{p_\gamma + p_\tau - 1}{p_s - p_\gamma} \right)}$$

$$\times \frac{1}{\left[1 + m_r \left[\dfrac{pf_s \displaystyle\sum_{i=1}^{q} \lambda_{s.i}}{n} + \Delta\lambda_{f(i-k)} + \cdots + \Delta\lambda_{f(i-2)} + \Delta\lambda_{f(i-1)} + \Delta\lambda_{b(i+1)} + \Delta\lambda_{b(i+2)} + \cdots + \Delta\lambda_{b(i+n)} + \lambda_{bf} + \lambda_c + \lambda_{tr} \right] \right]} \tag{5.32}$$

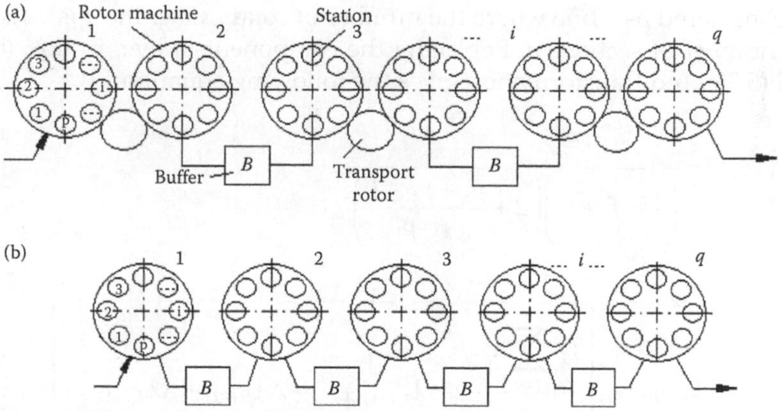

Figure 5.25 Scheme of a rotor-type automated line segmented on sections with buffers (a) and a rotor-type automated line with buffers after each rotor machine (b).

Equation of productivity rate for the rotor-type automated line with buffers of big capacity is represented as follows:

$$Q = \frac{p}{\left(\dfrac{t_{mo}}{q} f_c + t_a\right)\left(2 + \dfrac{p_\gamma + p_\tau - 1}{p_s - p_\gamma}\right)} \times \frac{1}{\left[1 + m_r \dfrac{pf_s \displaystyle\sum_{i=1}^{q} \lambda_{s.i}}{n} + \lambda_{bf} + \lambda_c + \lambda_{tr}\right]} \quad (5.33)$$

where all parameters are as specified earlier.

Equations (5.32) and (5.33) are similar to Eqs. (5.13) and (5.14) that enable solving mathematical task of optimisation of the structure for the rotor-type automated line segmented on sections with buffers, i.e. finding the optimal number of serial rotary machines q. Solutions of Eqs. (5.32) and (5.33) for optimisation of optimal number of serial workstations q with the number of sections n is defined by numerical decision. The task of the optimal number of serial rotary machines is solved by the first derivative of Eq. (5.33) when failure rates of serial workstations for all rotor-type machines are equal. However, practically this assumption is not real.

The mathematical model of productivity rate for the rotor-type automated lines with embedded buffers after each rotary machine is derived from analytical consideration presented for the section-based automated line of parallel–serial structure with buffers embedded after each automated line of parallel structure (Eqs. 5.15 and 5.16) and for the rotor-type automated line segmented on sections, Eqs. (5.32) and (5.33). Hence, the rotor-type automated line with buffers after each rotary machine should be considered as a line where the number of rotary machine q is equal to the number of sections n. Replacing the components n by q in Eqs. (5.32) and (5.33), and transformation gives the following equation:

$$Q = \frac{p}{\left(\dfrac{t_{mo}}{q} f_c + t_a\right)\left(2 + \dfrac{p_\gamma + p_\tau - 1}{p_s - p_\gamma}\right)}$$

$$\times \frac{1}{\left[1 + m_r \dfrac{pf_s \displaystyle\sum_{i=1}^{q} \lambda_{s.i}}{q} + \Delta\lambda_{f(i-k)} + \cdots + \Delta\lambda_{f(i-2)} + \Delta\lambda_{f(i-1)} + \Delta\lambda_{b(i+1)} + \Delta\lambda_{b(i+2)} + \cdots + \Delta\lambda_{b(i+n)} + \lambda_{bf} + \lambda_c + \lambda_{tr}\right]} \quad (5.34)$$

$$Q = \frac{p}{\left(\dfrac{t_{mo}}{q} f_c + t_a\right)\left(2 + \dfrac{p_\gamma + p_\tau - 1}{p_s - p_\gamma}\right)} \times \frac{1}{1 + m_r\left[\dfrac{pf_s \displaystyle\sum_{i=1}^{q} \lambda_{s.i}}{q} + \lambda_{bf} + \lambda_c + \lambda_{tr}\right]}$$

(5.35)

where Eqs. (5.34) and (5.35) are for buffers of limited and big capacity, respectively, and other parameters are as specified earlier.

Referring to the obtained result, the working examples for solutions of the number of serial workstation for the automated line of parallel–serial structure with buffers after each automated line of parallel structure can be used for the rotor-type automated line with buffers after each rotary machines.

Equations (5.34) and (5.35) are similar to Eqs. (5.15) and (5.16) for the productivity rate of an automated line of parallel structure arranged in the serial structure with buffers after each workstation. The solutions for productivity rates are similar and do not give optimal decision for the number of serial workstations. Equations (5.34) and (5.35) give an asymptotic increasing in productivity rate to defined limits that depends on the minimal machining time t_{min}, which enables the decomposition of the technological process.

5.4 Comparative analysis of productivity rates for manufacturing systems of parallel–serial arrangements

The manufacturing systems of parallel–serial arrangements are most complex in design and productive relative to other constructions. The area of application of these systems is also wide and includes manufacturing of work parts with complex design implemented by the labour-consuming technological processes and simple work parts with short technology of machining. Manufacturers prefer such systems, but complexity in their designs for machining the work parts of big sizes restricts this intention. This is the reason that automated lines of parallel–serial structures are designed with 2–3 parallel lines. The manufacture of the work parts or products with simple technological processes is implemented on the rotor-type machines arranged into serial structures with small number of serial structures. Such great variety in design of manufacturing systems of parallel–serial structure gives the same variety of productivity rates. This diversity in designs and productivity rates can be evaluated by comparative analysis of productivity rate.

The mathematical models for the productivity rate of manufacturing systems of parallel–serial structure are represented in Sections 5.1–5.3. These mathematical models for computing the values of productivity rates are used for the following analysis. The comparative analysis of productivity rates is conducted for the defined parallel–serial structures of manufacturing systems with one technological process, one number of parallel and serial workstations. For simplicity of computing, the productivity rate for the manufacturing systems of parallel–serial structure and for the same structure of automated lines that segmented into sections accepted the following properties: the buffers possess big capacity and the failure rate of workstations is identical. The mathematical models for productivity rate of the rotor-type automated lines are omitted from comparative analysis, because their productivity rates are almost twice less than for an automated line of parallel–serial structure. This assumption does not change the common results in differences of the productivity rates for the systems of parallel–serial structures. The initial data of manufacturing systems of parallel–serial structures is represented in Table 5.13. For simplicity of computing, the equal reliability of the serial workstations is accepted. Substituting initial data into equations of productivity rates of manufacturing systems of parallel–serial structure and computing yield the values of productivity rates (Table 5.14).

Figure 5.26 demonstrates the results of productivity rate for manufacturing systems of parallel–serial arrangement where abscissa represents the following: (1) manufacturing system of parallel–serial arrangement with an independent work of machines; (2) an automated line of parallel structure with independent serial constructions; (3) an automated line of serial structure with independent serial constructions; (4) an automated line of parallel–serial structure; (5) automated line of parallel–serial

Table 5.13 Technological and technical data of manufacturing systems of parallel–serial arrangement

Title	Data
Machining time, t_{mo} (min)	5.0
Auxiliary time, t_a (min)	0.2
Failure rate of workstation, λ_s	1.25×10^{-2}
Failure rate of transport system, λ_{tr}	4.0×10^{-4}
Failure rate of control system, λ_{cs}	6×10^{-8}
Failure rate of buffer, λ_b	6×10^{-8}
Mean repair time, m_r (min)	2.5
Number of serial workstations, q	6
Number of parallel workstations, p_s	4
Number of sections, n	3

Table 5.14 Productivity rate of manufacturing systems of parallel–serial structure

No	Manufacturing system	Equation of productivity rate	Productivity rate, Q (work parts/min)
1	Independent workstations	$Q = \dfrac{p}{(t_{mo}/q)+t_a} \times \dfrac{1}{1+m_r\lambda_{s.b}}$	3.753
2	Automated line of parallel structure with independent serial constructions	$Q = \dfrac{p}{(t_{mo}/q)+t_a} \times \dfrac{1}{1+m_r(p_s\lambda_{s.b}+\lambda_{cs})}$	3.440
3	Automated line of serial structure with independent parallel constructions	$Q_3 = \dfrac{p}{(t_{mo}/q)+t_a} \times \dfrac{1}{1+m_r\left(\displaystyle\sum_{i=1}^{q}\lambda_{s.i}+\lambda_{tr}+\lambda_{cs}\right)}$	3.539
4	Automated line of parallel–serial structure	$Q_3 = \dfrac{p}{(t_{mo}/q)+t_a} \times \dfrac{1}{1+m_r\left(p_s\displaystyle\sum_{i=1}^{q}\lambda_{s.i}+\lambda_{tr}+\lambda_{cs}\right)}$	2.212
5	Automated line of parallel–serial structure segmented on sections with buffers	$Q_3 = \dfrac{p}{(t_{mo}/q)+t_a} \times \dfrac{1}{1+m_r\left[\left(p_s\displaystyle\sum_{i=1}^{q}\lambda_{s.i}\right)\Big/n+\lambda_b+\lambda_{tr}+\lambda_{cs}\right]}$	3.097
6	Automated line of parallel–serial structure with buffers after each serial construction	$Q_3 = \dfrac{p}{(t_{mo}/q)+t_a} \times \dfrac{1}{1+m_r\left[\left(p_s\displaystyle\sum_{i=1}^{q}\lambda_{s.i}\right)\Big/q+\lambda_b+\lambda_{tr}+\lambda_{cs}\right]}$	3.440

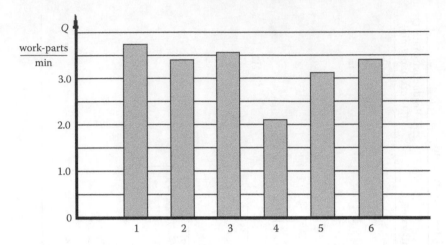

Figure 5.26 Productivity rate in histogram of manufacturing systems of parallel–serial arrangement.

structure segmented on sections with buffers; (6) automated line of parallel–serial structure with buffers after each serial construction. The values of productivity rate for manufacturing systems of parallel–serial structure are represented and observed clearly through the histogram with bars.

The comparative analysis of the productivity rates for manufacturing systems of parallel–serial arrangement demonstrates the following. The highest productivity rate gives the manufacturing system of parallel–serial structure with an independent work of workstations (1) and the lowest rate gives the automated line of parallel–serial structure with hard joining of workstations (4). The automated line of parallel–serial structure segmented on sections with buffers enables increase in the productivity rate (5). Last design of the automated line of parallel–serial structure with buffers after each parallel construction gives the productivity rate (6) a little less than the manufacturing systems of parallel–serial structure with an independent work of serial constructions (2). The manufacturing systems of parallel–serial structure with an independent work of parallel constructions (3) give bigger productivity rate than an independent work of serial constructions (2). All manufacturing systems of parallel–serial arrangement are characterised by their own properties. Arrangements number (6), (5), (1), (2), (3) and (4) occupies production area from maximal to minimal and cost from minimal to maximal, respectively, due to design principles. Economics principles are a final step for evaluation of manufacturing systems design that is preferred for production systems.

Bibliography

1. Altintas, Y. 2012. *Manufacturing Automation.* 2nd ed., Cambridge University Press. London.
2. Badiru, A.B., and Omitaomu, O.A. 2011. *Handbook of Industrial Engineering Equations, Formulas, and Calculations.* Taylor & Francis. New York.
3. Benhabib, B. 2005. *Manufacturing: Design, Production, Automation, and Integration.* 1st ed., Taylor & Francis. New York.
4. Ben-Daya, M., Duffuaa, S., and Raouf, A. 2000. *Maintenance, Modeling, and Optimization.* Kluwer Academic Publishers. New York. USDA.
5. Chryssolouris, G. 2006. *Manufacturing Systems: Theory and Practice.* 2nd ed., Springer. New York.
6. Groover, M.P. 2013. *Fundamentals of Modern Manufacturing: Materials, Processes, and Systems.* 5th ed., (Lehigh University). John Wiley & Sons. Hoboken, NJ.
7. Nof, S.Y. 2009. *Springer Handbook of Automation.* Purdue University. West Lafayette, IN.
8. Rao, R.V. 2011. *Advanced Modeling and Optimization of Manufacturing Processes.* 1st ed., Springer Series in Advanced Manufacturing. New York.
9. Shaumian, G.A. 1973. *Complex Automation of Production Processes.* Mashinostroenie. Moscow.
10. Shell, R.L., and Hall, E.L. 2000. *Handbook of Industrial Automation.* Marcel Dekker. Inc. New York.
11. Usubamatov, R., Sin, T.C., and Ahmad, R. 2016. Mathematical models for productivity of automated lines with different failure rates for stations and mechanisms. *The International Journal of Advanced Manufacturing Technology.* DOI 10.1007/s00170-015-7005-6.
12. Usubamatov, R., Ismail, K.A., and Shah, J.M. 2012. Mathematical models for productivity and availability of automated lines. *International Journal of Advanced Manufacturing Technology.* DOI 10.1007/s00170-012-4305-y.
13. Usubamatov, R., and Sartov, T., 2017. A mathematical model for productivity rate of APSL segmented on sections with buffers of limited capacity. *The International Journal of Advanced Manufacturing Technology.* DOI 10.1007/s00170-017-0442-7.
14. Usubamatov, R., Abdulmuin, Z., Nor, A., and Murad, M.N., 2008. Productivity rate of rotor-type automated lines and optimization of their structure. *Proceedings of the Institution of Mechanical Engineers, Part B: Journal of Engineering Manufacture.* 222(11). pp. 1561–1566.
15. Volchkevich, L.I. 2005. *Automation of Production Processes.* Mashinostroenie. Moscow.

chapter six

Analysis of the work of industrial machines and systems at real manufacturing environment

The manufacturing industry has to face various challenges to stay competitive in the market place. Therefore, industrial managers are striving to increase productivity and use various techniques and approaches. There are different approaches in the area of engineering processes considering all aspects of the manufacturing systems, which enable to solve the main problems of companies. Manufacturers bring the experience of Kaizen, Taguchi and handbooks with practical solutions of manufacturing processes of products and operations management. However, not many methodologies have carried out the work on the holistic principles and dealt with the use of mathematical model based on technology of production processes, and reliability of mechanisms, machines and management, which reflected on productivity. Method for analysis of the work of industrial machines and systems at real manufacturing environment enables to improve production process. The main purpose of this method set is to overcome the problem for real application of mathematical model of productivity rate in industry. Engineers consider design for manufacturing systems, components, processes, performance and comparative parameters of technology and reliability of production systems. The mathematical model expresses this dependency, where availability of manufacturing system have right links to productivity rate of the system and the number of technicians that serve these systems.

6.1 Introduction

The manufacturing industry has to face various challenges to stay competitive in the market place. Therefore, industrialists are striving to increase productivity and use various techniques and approaches. Modern production processes are carried out using complex manufacturing systems. Industrial managers are looking for ways to use all equipment in full capacity and to obtain maximum productivity and efficiency. In reality, in the field of complex system, especially when it involved the exploitation of production lines, these systems do not operate at their maximum capacity due to many reasons. Extensive analysis on established manufacturing processes shows that production lines do not work constantly. There are times when the production lines do not work and the productivity rate demonstrated by these systems is less than calculated by the operating manuals.

Various studies have been conducted in the past to improve the manufacturing processes of productivity. There are different approaches in the area of engineering processes considering all aspects of manufacturing systems, which enable to solve the main problems of companies. Engineers consider design for automated and assembly processes and enumerate the components, processes, performance and comparative economics of several types of production systems. Manufacturers bring the experience of Kaizen, Taguchi and others, explains how to implement waste reduction, time and motion studies, line balancing, quality-at-the-source, visual management and production line design. There are handbooks with practical solutions of manufacturing processes of products and operations management, which explain the ideas and concepts and outlines the procedures and techniques involved.

Manufacturers consider problems of efficient production line design, which depend on technical personnel working in design, planning and production departments in industry as well as managers in industry. There are guides to practitioners and researchers involved in the planning, control and management of manufacturing systems, which include a wide range of available methodologies and tools. The methodology of time series analyses gives a statistically significant adverse impact on total labour hours, overhead hours and downtime in the production line, minor repair and major rework. However, not many methodologies have carried out the work on the holistic productivity rate and dealt with the use of mathematical model based on technology of production processes, and reliability of mechanisms, machines and management, which reflected on productivity of industrial processes. Manufacturers need some holistic methods that to increase the productivity rate of production systems and enable calculating the maximum productivity rate, types and reasons of productivity losses of real production processes. To solve the actual problems of production processes, many practical methods for analysis of the work for current manufacturing systems were developed, which always have bottleneck components. Among known practical methods for analysis of production processes, the interest due to holistic approach for solving of the production problems is presented [15,21].

The system analysis of production processes, which aim is to define, calculate and demonstrate directions for the engineering solutions to improve production processes is presented as the powerful and effective tool. This system of analysis enables for solving the problems of productivity enhancement and efficiency of complex manufacturing systems some times without big expenses on practical studies of production processes. The results of the holistic analysis are verified using mathematical approaches. This system analysis of production processes is a universal method and can be applied to any kinds of production processes.

The time taken (observation time) for the analysis of manufacturing systems is classified as working process time and idle time. Manufacturing process time can be considered from actual work time and auxiliary time. Work time is the spent on machining or assembly of work parts and units using different tools. Auxiliary time is spent on preparation of machining or assembly process that includes delivering work parts and units to machining or assembly area, delivering tools, preparing other components for manufacturing process, etc. Idle time is the lost time, because a production system does not produce product and causes the reduction in company productivity and factory incomes. Generally, the cause of idle times is technical, managerial and organisational reasons like reliability of machinery, absence of work parts to be manufactured, drop of energy supply, low discipline of employees, etc.

These idle times of manufacturing process and decrease of the productivity rate of the system can be represented as a mathematical expression. In manufacturing engineering, the term productivity rate is formulated as the number of products manufactured per some observation time. The current popular methods to express productivity rate in industry are presented using the following terms:

- Cycle time is the period required to complete one cycle of an operation or to complete a function, job or task from start to finish. Cycle time is used to specify the total duration of a process from start to finish, whereas its terms have limitation because of its inability to express the actual productivity rate of the entire system.
- Tact time is defined as the maximum time per unit allowed, producing a product to meet demand. Tact time sets the pace for industrial manufacturing lines. Tact time cannot measure the losses of productivity rate.
- Productivity rate is the number of products fabricated per observation time with dimensions of products in discrete work parts, length, volume, weight, etc.
- Productivity is a ratio of output/input and used in macroeconomics for evaluation of the efficiency of system. For evaluation of the productivity rate of a machine and a production line, this term is not an appropriate index.

However, industrial environment is more complex, and the productivity rate of manufacturing systems depends on technology of processes, reliability of machines and mechanisms, managerial and organisational problems. Hence, it is important to derive the mathematical model regarding industrial factors on productivity rate of manufacturing systems. Different manufacturing systems in electronic, machining, transport, textile, chemical, etc. industries have different factors that lead to downtimes of production processes. These particularities should be presented in methods for evaluation of productivity of manufacturing systems. Analysis of the typical manufacturing processes in the machining industry demonstrates the following factors that are reasons of productivity rate losses:

1. Cyclic time losses: auxiliary time for manufacturing products that is part of cycle time.
2. Non-cyclic time losses that are sum of random downtimes of production system by the following reasons:
   - Reliability cutters and tools.
   - Reliability of machines, mechanisms, devices and electrical and electronic components.
   - Manufacturing defected products.
   - Managerial and organisation problems as absent of work parts, drop of power, low discipline of employees, reorganisation of production processes, etc.

Presented factors are not full and can be added, while some others are inherent in machining processes. In case of other type of industries, some factors can be omitted, but added to the other one. For example, transport industries do not have cutters and do not produce defected products, but have other problems like traffic jam, rush hour, etc. These particularities of industries lead to include into methods of own factors for evaluation of the efficiency of industrial systems.

6.2 A method for assessing productivity rate of manufacturing system at real industrial environment

Mathematical and statistical analysis method is required to apply for productivity rate in industry for visually overviewing the failure factors and further improvement within the production line, especially for manufacturing lines with complex structure. Since this approach is complicated, the concept of engineering mathematical analysis method is used to analyse the productivity rate of manufacturing system in real industrial environment. Manufacturing researchers, engineers and practitioners developed the analytical method for evaluation of the production processes efficiency. This holistic method for assessing productivity rate based on statistical analysis enclosed all components of the manufacturing process [15,21]. Practical application of this method is represented later [16].

Mathematical analysis method and validation purpose focused on selection and design the most appropriate methods to obtain better and high trustworthiness result from a company. Engineering method considers a set of method guideline to solve engineering problem with support of engineering and mathematic theory generally. A set of engineering mathematical analysis method designed specifically to solve the problem for real application of complicated mathematical model of productivity rate. This method consists of four stages data collection, calculation and comparison, analysis and sustainable improvement which are summarised as DCAS.

The main purpose of this method set is to overcome the problem of real application of mathematical model of productivity rate in industry. This problem can be solved by applying and following DCAS method, which contains all important ways, methods, analysis and improvement to apply a mathematical model. Each of the stages of methods is discussed clearly as follows.

Data Collection. The first stage of the engineering mathematical analysis method is data collection, since it is necessary to obtain a high quantity of trustworthy data for computing productivity rate and failure rate analysis of a manufacturing equipment. Engineering is required to apply the data collection method, in which observation and interview are done to collect and record the important data. However, there will be a large amount of data to review and record. Data collection method is implemented according to the retrospective and observational study. The historical data of pass months for the production line are collected, analysed and computed for the average number of products manufactured per shift. To achieve the statistical requirement, several months of data are recorded together with the daily finish products manufactured. Industrial practice demonstrates that the number of products manufactured per production shift is not stable and can have lot of fluctuations. The final finish product per shift produced varies in every shift as different failures occurr. To achieve the reliability of study for obtaining the sufficient number of shift, the data are represented in the diagram 'average mean cumulative number of products manufactured per shift versus the number of shifts'.

Based on the diagram of average mean cumulative number of products manufactured per shift, a researcher can define the average mean cumulative products that falls in the tolerance level of ±5% from average number of products manufactured per shift. This number of products defines the length of research observation time. Further additional shifts of observation will not change the average cumulative number of parts produced per shift. An observation time of defined shifts is enough to get 90% statistical trustworthiness of the average research results. These results will be used in the following mathematical calculations of productivity rate.

Chronometric analysis table can help to simplify and summarise the main parameters and time observation data. By referring to the literature review or analytical approached for data reliability, the number of data should be set up by rules of mathematical statistics and probability theory, which can achieve 90% confidence level. After that, there will be two standard engineering method tables for data collection, such as technical data table and reliability indices table. Both the tables contain important parameters used in mathematical model calculation.

Based on the data of two tables, engineers have to take the readings and record the data that are required. The failure rate of workstations depends on the number of machines being used for the process in productivity. The total machining time is the sum of all durations for every station to process. There are three categories of data required to collect for this method: actual production data, technical data and reliability data. Details of all data are shown clearly in Table 6.1.

Table 6.1 Overall detail data list required for engineering mathematical analysis method

Data category	Collection method	Data parameter	Symbol	Unit
Actual productivity rate	Retrospective study	Number of products manufactured	z	product
		Time used	θ	min
Technical data	Observational study	Total machining time	t_{mo}	min
		Auxiliary time	t_a	min
		Number of workstations	q	
		Correction factor of bottleneck machining time	f_c	
Reliability data	Retrospective study	Workstations failure rate	$\lambda_{.si}$	1/min
		Control system failure rate	$\lambda_{.c}$	1/min
		Transport system failure rate	$\lambda_{.tr}$	1/min
		Defected parts failure rate	λ_d	1/min
		Management problem failure rate	λ_d	1/min
		Logistic problem failure rate	λ_d	1/min
		Mistaken assembly failure rate	λ_d	1/min
		Mean time to repair	m_r	min

The first category, which is the actual productivity data, is the data that contain the number of work parts produced and time using retrospective study. Retrospective study is a data collection method that refers to previous and near past data. Number of work parts produced and time used are required to measure and calculate the actual productivity rate in an automated line, which represents as guideline for mathematical model comparison use.

Technical data is a set of data regarding the technical work in an automated line such as machining time, auxiliary time and idle time. The number of workstations and correction factor of bottleneck machining time are also included in technical data set due to both parameters that affect the productivity rate. These category parameters are improved via technical work. Observational study methods such as time study or chronometric analysis, interview and focus group are applied to obtain high confidence level data set.

Retrospective study is applied to collect reliability data such as failure rate of workstations, control system, transport system and defected work parts, which are related to the reliability parameters specific in failure rate of an automated line. Mean time to repair is also considered and calculated through the total failure time divided by total number of failures per observation time. For retrospective study data, at least 90% of data trustworthiness is required for engineering data. The deviation of 10% is calculated due to the total of 5% from the mean value and it is calculated using the following equation:

$$\delta_i = \frac{\sum_{i=1}^{n} z_i - z_{av}}{z_{av}} \tag{6.1}$$

Derivation is represented in Eq. (6.1), which is the yield of number of work parts produced in i shift N_i minus average or mean of work parts produce in i shift z_{av} and divided by mean z_{av}. For example, average or mean of work parts produce in i shift z_{av} is 40 work parts. If 90% of statistical trustworthiness is reached, the deviation δ_i is required in range of 5%. For this 5%, is become two limits for upper and lower deviations. Consequently, to reach 90% of statistical trustworthiness, the value z_i is required to convert average mean cumulative number since the production of work parts is usually highly fluctuated to reach the range of 40± work parts. Since the data collection included failure rate parameter, 160–170 times of failure is required to obtain 90% of probability trustworthiness according to the rules of mathematical statistics. So, to obtain 90% of statistical and probability trustworthiness, the retrospective data should consist of 160–170 events of failures or breakdowns of automated flow line and 5% deviation of work parts quantity produced.

Calculation and Comparison. The second stage of engineering mathematical analysis method is to calculate the data collected using mathematical model specified earlier and compare the yield of the calculation. Calculation will be carried out manually or using the computer software (Mathcad, MATLAB) as recommended, primarily intended for the verification, validation, documentation and re-use of engineering calculations. This software is used by engineers and scientists in various disciplines – most often those of mechanical, chemical, electrical and civil engineering. The calculation is performed based on the mathematical model for an automated line that is defined as the bottleneck for production system.

Actual productivity data set is used to calculate the real productivity rate of work parts by using the equations represented in Chapters 2–5. Both sets of technical and reliability data are applied in equations of Chapters 2–5, which contains mathematical models for productivity rate of automated line with average and different indices of workstation's failure rates.

Comparison of data and results is necessary to compare the differences between the newly developed mathematical model output with current model and actual productivity output. There are two types of graphical methods chosen for analysis and the differences for the results are compared. Line chart with markers and histogram is the best way to represent the results and data analysis. The comparison of results computed by proposed mathematical model of productivity rate to the actual productivity rate in an automated line demonstrates differences in results. The purpose of compare is the validation of the mathematical model for productivity rate.

Analysis for Failure Rate. The analysis for failure rate is a necessary step, where the factors of time, downtimes, and hence productivity losses and failure can be determined. The practical study of failure rates and reasons enable finding out the unpredicted parameters, which does not consider the mathematical model of productivity rate. The determination of losses of the problems can be observed and improvement can be done based on the largest amount of losses in productivity. The losses of productivity rate for a manufacturing system in the actual condition can be demonstrated in the losses productivity diagrams. In analysis of failure rate stage, the productivity rate losses diagram is the powerful tool for sustainable improvement of a manufacturing system. This analysis specifically presents the losses of productivity rate due to different of failure reasons graphically and visually using derived equation and mathematical computed results. For detailed analysis, the failures of manufacturing system should be separated on type of their nature as mechanical, electric, electronic, managerial, etc. This separation is necessary for the following study and improvement by corresponding engineers. Through analysis, the impact of causing losses of productivity rate in manufacturing system, for which the failure reasons are detected and represented by their values.

In addition, Pareto diagram analysis can be applied to arrange and analyse the reason of causing losses of productivity rate from highest to lowest impact parameter.

Sustainable Improvement. In the final stage of engineering analysis, method is considered as the sustainable improvement of production system. The parameters obtained from analysis, which are different values of reliability of workstation, mechanisms, bottleneck time of workstation, rejected work parts, managerial problems, etc., will be categorised into four matrices. Since there are many reasons causing the losses of productivity rate in a manufacturing system, the Priority, Action, Consider, Eliminate (PACE) Prioritisation Matrix is introduced to select which parameters are preferred to improve in the first place, which one in the second turn and which are not suggested to improve. There are four levels in the PACE Prioritisation Matrix. Priority group parameters are preferred to proceed for improvement since they have a high impact to decrease the productivity rate loss and more easy to overcome. After priority group, the preferable group for further improvement to increase the productivity rate of a manufacturing system is action group and followed by consider group and last but not least is the eliminate group. The group discussion is done in a closed group and is executed by a solution of engineers.

A Prioritisation Matrix assists engineers what to do after key actions have been identified, but their relative priority is not known with certainty. Prioritisation matrices are especially useful if problem-solving resources for production system are limited, or if the identified problem-solving actions are strongly interrelated. To create a matrix, it is necessary to identify relative ability of each possible action to effectively deliver the results for sustainable improvement of a production system. The results of computing the productivity rate and losses are a weighted ranking of all the possible actions that require consideration. The finished matrix assists a team of engineers and managers to make an overall decision and determine the sequence in realisation of results toward an objective.

General Approach. Study and analysis of productivity rate in a production system environment consists of the following purpose:

- To determine the productivity rate losses and reasons.
- To determine potential reserves to enhance productivity rate of the production system in industrial environment.
- To provide initial parameter for planning a new production system based on general experience from actual operating arrangement.
- To determine management approaches for a production system to enhance productivity.

The calculation of the productivity rate by equations presented in previous chapters can be applied for various types of production systems (manual, automatic). The complexity of calculation depends on the production processes, which can be based on technological processes and structure of production systems. Therefore, formalisation of productivity calculation must cover all its complexity mathematically. The result of calculation should easily transform sufficient readable diagram to provide reliable information for corporate decision makers.

Manufacturers would like to get detailed information regarding the downtimes of production systems, reasons and numerical data of productivity losses. The nature of the downtimes and time losses of production processes is different and random. The proposed method gives information about the real productivity and values of its losses for production system. This method is developed on the base of real statistical data as result of observation of the work process of production system. The following analysis and mathematical calculation enable to demonstrate various productivity losses of manufacturing systems, reasons and volume one. The improvement of the work of production system can be made based on its productivity losses. Study and analysis of production processes included the following components [15,21]:

- To study the production process of manufacturing product (methods, routes and other technical and technological requirements).
- To study the construction of all machines, stations and mechanisms involved in production process.
- To study the reasons of machines, stations and mechanisms failures.
- To study the organisational and managerial aspects of the production process.

The study and analysis of production process is conducted around two decades of production shifts by fixing the actual work time and by recording the time and reasons of failures of all machines and mechanisms for the production system. The number of production shifts can be different in different industries. It depends on the stability of production processes. The statistical records of production system work per observation time are included: the time spent in manufacturing of products on each machine or workstations, the time spent in idle conditions that consist of technical, organisational and managerial problems, methods of repair, reasons of downtimes, number of products produced per shift and the duration of each station cycle. The duration of production processes and cycle time are calculated in average magnitudes. It is necessary to define the maximum duration of the production cycle time (bottleneck station) that defines the productivity rate per cycle time of the manufacturing system. These records contain all information required to characterise manufacturing processes of the product.

During observation of the manufacturing system, the volume of data collection should satisfy the necessary requirements of mathematical statistic rules and regulations. To get reliable results of studying and researching, the manufacturing system output used company's statistical data to monitor previous several months of the system work. The obtained data are used to calculate the average productivity rate per shift N_{av}, which is basis for the following statistical calculations. The average deviation of the productivity rate for the time of observation should be less than ±5% from accepted an average basis. Hence, the deviation of the average mean cumulative number of manufactured products should be in the tolerance.

The method presented earlier for evaluation of the productivity of the manufacturing system is universal and can be applied for any type of process, i.e. it can be automatic, semi-automatic or manual like most assembly processes. Employees with tools of minimum automation conduct assembly processes of most manufacturing systems manually. Assembly processes are calculated by technologist and managers and fulfilled at prescribed average time of manufacturing processes. Hence, manual operations can be described in terms of productivity, which are cycle, auxiliary times and time losses, by different reasons, etc.

The actual parameters of productivity of any manufacturing system is expressed by the equations presented in Chapters 2–5. These equations can be used for assembly process. The work time is considered as the assembly time. Then, the obtained ratio of the total work time divided by the number of product assembled $\theta_w/z = T$ is logically presented as the cycle time T of the assembly process for one product. The cycle time T of assembly process is considered as a sum of total work time that expresses right assembly process t_w and auxiliary time t_a that expresses the time assembly spent on preparatory motions before and after process, so $T = t_w + t_a$.

The total idle time θ_i of the assembly line is considered as time losses in time observed and may be caused by different reasons like malfunctioning of the conveyor and tools, power drop, defect parts, scheduling problems, etc. Hence, the idle time is the sum of any non-productive time relatively to the number of parts assembled during the process in production line and formulated by the expression presented in Chapter 2.

As mentioned earlier, the reasons of the idle time can have different natures of origin. In an industry, it is necessary to separate the reasons of the idle time on two groups: technical reasons t_{tech} and organisational and managerial reasons t_{org}. Both of them reflect two types of production activities and enable to evaluate the efficiency of them.

Finally, actual productivity of assembly process is represented by same equation [18] as for machining processes but with symbols of assembly process. However, there is difference in the equation of productivity rate. The auxiliary time in assembly processes is part of assembly time and depends on the number of assembly stations. For the segmented assembly process on q serial stations, the average auxiliary time on each station is reduced proportional to the number of stations (t_a/q). The total auxiliary time in assembly process is presented as the sum of times spent on delivery of the machine parts to an assembly area and delivery of the tools, accessories, etc. for an assembly process and move away from an assembly area. Usually, the time spent on each such motion is short and evaluated by several seconds. In industry, this time is standardised for the calculation of cost of an assembly process.

Increase in the number of serial stations leads to an increase in the number of employees, which use tools and mechanisms for assembly process. These changes can be mathematically expressed in equation of the productivity rate for assembly line and presented by the analytical model based on previous studies represented in Chapters 2–5.

$$Q = \frac{1}{\dfrac{t_w + t_a}{q} f_c + q \dfrac{\theta_{st}}{z} + \dfrac{\theta_c}{z} + \dfrac{\theta_{org}}{z}} \tag{6.2}$$

where $(t_w + t_a)/q$ is the average assembly cycle time of one station; $q(\theta_{st}/z)$ is the idle time due to failures of tools and mechanisms of one station that lead to stops of all stations in the assembly serial line; θ_c/z is the time referred to one product due to failures of common mechanisms that do not depend on changes of number of stations (central air compressor, central electrical power supply, etc.); θ_{org}/z is the time referred to one product due to organisational and managerial problems; other parameters are as specified earlier. The time loss due to managerial and organisational problems does not change with the change of the number of line stations.

Equation (6.1) expresses the real productivity rate of the assembly line and analysis of the equation show that it has extreme value where the variable is the number of assembly stations q. The first derivative of Eq. (6.2) gives an optimal number of stations q by criterion of maximum productivity of an assembly line.

$$\frac{dQ}{dq} = \frac{d}{dq} \left(\frac{1}{\dfrac{t_w + t_a}{q} f_c + q \dfrac{\theta_{st}}{z} + \dfrac{\theta_c}{z} + \dfrac{\theta_{org}}{z}} \right)$$

giving rise to the following:

$$-\left(-\frac{t_w + t_a}{q^2} f_c + \frac{\theta_{st}}{z} \right) = 0$$

Then optimal number of serial stations in assembly line is given by the following equation:

$$q_{opt} = \sqrt{\frac{(t_w + t_a)f_c z}{\theta_{st}}}$$

(6.3)

where all parameters are specified earlier.

Substituting Eq. (6.2) into Eq. (6.1) and transformation yield the equation of the maximal productivity rate for the assembly line represented by the following expression:

$$Q_{max} = \frac{1}{\frac{(t_w + t_a)f_c}{\sqrt{(t_w + t_a)f_c z/\theta_{st}}} + \frac{\theta_{st}}{z}\sqrt{\frac{(t_w + t_a)f_c z}{\theta_{st}}} + \frac{\theta_c}{z} + \frac{\theta_{org}}{z}}$$

$$= \frac{1}{2\sqrt{\frac{(t_w + t_a)f_c z}{\theta_{st}}} + \frac{\theta_c}{z} + \frac{\theta_{org}}{z}}$$

(6.4)

Conducted analytical approach of the assembly serial line productivity and results of study of the car cabins assembly line enable the system analyst to calculate the productivity rate and optimal number of assembly stations of an assembly line. The practical application of the proposed method for evaluation of the productivity and losses for the assembly line as part of car manufacturing process is presented as follows.

6.2.1   Industrial case study

A car company is selected to apply and validate the mathematical model for productivity rate using engineering mathematical analysis method. Productivity rate and failure rate parameter causing loss of productivity in automated line in a car company are calculated and analysed based on DCAS method. The result for each stage of engineering mathematical method is represented separately in detail later. An investigation was carried out on the assembly process of a car cabin for one section of the production line with linear structure at the real car company (Figure 6.1) [19]. It was chosen one section of the assemble line, which result in the main bottleneck problem. The technical data of the assembly line are presented in Table 6.2.

One assembler is equipped with pneumatic tools for fulfilment of assembly processes services at each assembly serial station. The total assembly and auxiliary times of the car cabin are presented as a sum of

Figure 6.1 Car assembly line.

assembly and auxiliary times on each station (Table 6.2). Pneumatic tools are serviced by the centralised air compressor. The car cabin is located on wheeled pallets and is transported from one station to another using a conveyor. Schematically, this assembly serial line is represented in Figure 6.2.

Table 6.2 Main technical data of the car cabin assembly line

Title	Data
Number of assembly stations	$q = 18$
Number of employees	$n = 18$
Number of pneumatic tools	$f = 18$
Number of parts assembled in the line	$z = 350$
Centralised compressor	1
Average assembly cycle time	$T = 22.0\,\text{min}$
Average total assembly time	$t_w = 185.64\,\text{min}$
Average total auxiliary time	$t_a = 47.61\,\text{min}$

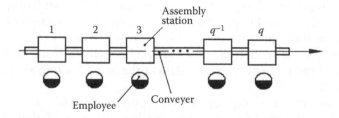

Figure 6.2 Scheme of the section of the car cabin assembly serial line.

The research was carried out for assembly serial line of the car cabin and all data were recorded according to the proposed methodology. After finishing the assembly process, the conveyor shifts the pallets one-step forward according to the technological process. In case of delay of the assembly process at some station due to technical and technological reasons, all other assembly stations stop until the assembly process finishes at the station delays. The sufficient number of shifts for the study and research and the number of cars assembled during the 24 production shifts per observation time are presented in Figure 6.3.

To evaluate the sufficiency of the number of shifts for the study and research, 'average mean cumulative number assembled car cabins versus number of shifts' is calculated and presented in Figure 6.4.

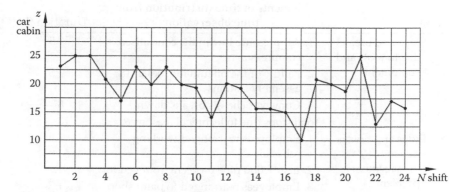

Figure 6.3 Number of car cabins assembled versus number of shifts.

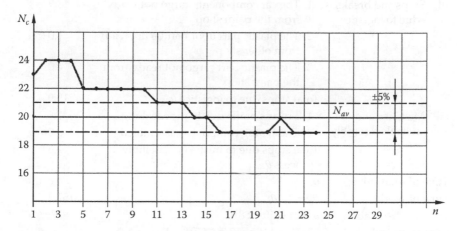

Figure 6.4 Average mean cumulative number of assembled car cabins versus number of shifts.

Figure 6.4 shows the average mean cumulative number of assembled car falls in the tolerance after 14–15 shifts of the time observation. Further additional shifts of observation will not change the average cumulative number of cars assembled per shift. Time observation of 14–15 shifts for this type of assembly process is enough to get 90% trustworthiness of statistically average results of research. These results will be used at the following mathematical calculations of productivity parameters of the assembly line.

The recorded information obtained from the study and analysis of assembly process of the car cabins is represented in Table 6.3. The obtained

Table 6.3 Chronometric analysis of the time observation of the car cabin serial line assembly process

Event		Elements of time distribution from time observation	Duration (h)
I	Stops and breaks due to tools and equipment	1. Air pressure low in the centralised compressor	1.5
		2. Failure of the pneumatic tool at 18 assembly stations	0.5
		3. Conveyor chassis breaks dawn	8
		Total stops due to technical reasons θ_t	10
II	Stops and breaks due to management problem	1. Employees rearranged working schedule	1.6
		2. Instruction pipe problem from vendor	1.0
		3. Employees rearranged to paint shop	1.7
		Total stops due to management problem θ_m	4.3
III	Stops and breaks due to logistics	1. The car components cargo not ready from the paint shop	10.9
		2. The plastic part door handle not ready from other shop	1.0
		3. The plastic part cargo not ready from the paint shop	4.0
		Total stops due to logistics θ_l	15.9
IV	Mistaken and not reliable assembly process	1. Process brake due to pipe leaking	1.6
		2. The wrong component assembled	3.0
		Total stops due to mistaken assembly process θ_{mis}	4.6
Time of total stops	θ_i		34.8
Time of work	$\theta_w = \theta - \theta_i$		157.2
Time of observation	$\theta = 24$ shifts $\times 8$ h		192
Cycle time, min	$T = (t_w + t_a)f_c = (17.3 + 4.7)1 = 22; f_c = 1.0$		
Number of car cabins assembled $z = 456$			

information enables the manufacturer to come up with the information on the capacity of the assembly line, the quality of assembly stations work, the provision to increase productivity and the reliability of the machine and mechanisms in use.

The balancing of the assembly process of the car cabin and cycle time distributions on assembly stations is demonstrated in Figure 6.5. The bottleneck assembly station determines the cycle time T for the assembly line that is used for computing its productivity parameters.

Reliability of the study and calculation results should be confirmed by comparing the work time obtained practically and theoretically.

Theoretical time of work of the assembly line is $\theta_{w.th} = Tz = 22.0 \, \text{min} \times 456 = 10{,}032 \, \text{min} = 167.2 \, \text{h}$, where T is the cycle time of the bottleneck assembly station and z is the number of products (Table 6.3). The deviation between practical θ_w and theoretical $\theta_{w.th}$ times is calculated by the following formula:

$$\delta = \frac{\theta_{w.th} - \theta_w}{\theta_{w.th}} \times 100\% = \frac{167.2 - 157.2}{167.2} \times 100\% = 5.98\%$$

The deviation of $\delta = 5.98\%$ is acceptable in mathematical statistics and should not exceed 10%, and the results of study can be considered trustworthy with probability of $\approx 94\%$.

The obtained information of the study enables to calculate technical data of the assembly line (Table 6.3). The main characteristics are determined from two types of documents, namely the actual assembly line cycle and the actual time observation spent. This information is used for calculating the availability and real productivity of the assembly line. To make the calculation more user friendly, it is necessary to go through it with reference to Table 6.3.

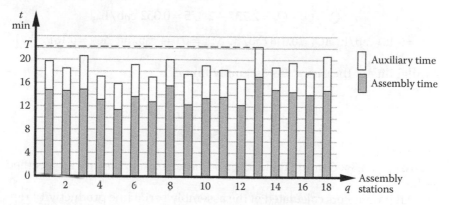

Figure 6.5 Cycle time distribution on stations of an assembly line.

Calculation of the basic parameters of the car cabins assembly line productivity is

- Limit or maximum of the productivity rate for the assembly line is

$$Q_l = \frac{1}{t_w} = \frac{1}{17.3} = 0.0578 \frac{cab}{min} = 3.468 \text{ cab/h.}$$

where $t_w = 17.3$ min is the maximum assembly time of the bottleneck station.

- Productivity rate per cycle time of the assembling flow line is

$$Q_c = \frac{1}{t_w + t_{aux}} = \frac{1}{22.0} = 0.0454 \frac{cab}{min} = 2.727 \text{ cab/h.}$$

where $T = (t_w + t_{aux}) = 22.0$ min is the maximum assembly cycle time of the bottleneck station, which defines the productivity rate of the assembly line.

- Actual productivity of the assembly serial line is

$$Q_a = z / \theta = 456 / 192 = 2.375 \text{ cab/h}$$

- Productivity losses of the assembly line due to defined reasons ΔQ_i are calculated on the basis of Table 6.3. Cyclic productivity losses due to auxiliary time are calculated by formula

$$\Delta Q_1 = Q_l - Q_c = 3.468 - 2.727 = 0.741 \text{ cab/h}$$

Other productivity losses of the assembly line are calculated by the ratios of idle times as $\theta_c : \theta_{mis} : \theta_m : \theta_l = 10 : 4.6 : 4.3 : 15.9$ (Table 6.3).

$$\Delta Q_i = Q_c - Q_a = 2.727 - 2.375 = 0.352 \text{ cab/h.}$$

$\Delta Q_2 = 0.101$ cab/h, $\Delta Q_3 = 0.046$ cab/h, $\Delta Q_4 = 0.044$ cab/hr, $\Delta Q_5 = 0.161$ cab/h.

Availability of the car cabins assembly line is

$$A = \frac{1}{1 + \dfrac{\theta_i}{zT}} = \frac{1}{1 + \dfrac{34.8 \times 60}{456 \times 22.0}} = 0.827$$

where $\theta_i = 34.8 \times 60$ min (Table 6.3), and other parameters are as specified earlier.

All parameters calculated of the assembly serial line productivity represented in Pareto chart (Figure 6.6) enable a system analyst to consider and analyse the change in the productivity rate due to different technical reasons and develop recommendation to decrease one.

Based on productivity losses analysis calculated earlier, Figure 6.6 is illustrating and summarising the results in graphical method. Pareto chart shows a decrease of productivity rate from potential productivity outputs until the actual outputs due to failure rates of assembly line units and mechanisms and other idle times of different reasons. The new mathematical model is having a close relationship to the actual outputs, where the markers are almost at the same level of position. The amount of productivity rate loss can be observed clearly through histograms with bars in grey colour. The losses of productivity rate are rearranged accordingly in descending order in Pareto chart (Figure 6.7).

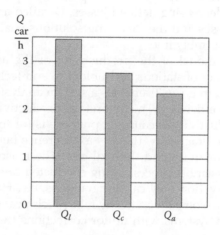

Figure 6.6 Productivity diagram of parameters Q_l, Q_c, Q_a of the car cabin assembly serial line.

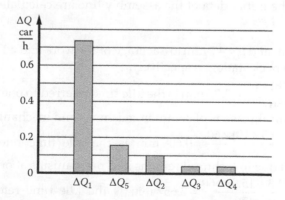

Figure 6.7 Arrangement of losses of productivity rate in Pareto diagram.

Pareto chart helps to analyse the data easily and provide a clearer view of results and priority directions for sustainable improvement of the assembly line of car cabin.

Graphical presentation of the assembly line productivity losses is shown in Figures 6.6–6.8, where major losses are due to auxiliary time ($\Delta Q_1 = 0.74$ cab/h), then logistic ($\Delta Q_5 = 0.16$ cab/h) and reliability ($\Delta Q_2 = 0.1$ cab/h), and problems are sensitive. The values of productivity losses by other reasons are relatively less. The figure is actually a summary of the calculation performed earlier. It means if the assembly line is 100% efficient, it can produce 3.46 car cabins/h. However, due to some losses during the process, it actually produces only 2.37 car cabins/h or the assembly line has 70% of its efficiency.

All these losses are the sources of productivity incremental in the assembly line by decreasing defined losses. Detailed analyses of reasons of productivity losses and the following solutions enable to increase the efficiency of the assembly line.

Mathematical model of the productivity of the assembly line as a function of a number of stations, technological and technical parameters, which is presented in Eq. (6.2), enables a system analyst to analyse all factors that influence efficiency. The theoretical productivity is not the same as the actual productivity because its purpose is for forecasting the system's productivity. Practically, there is a difference between mathematical model of productivity and the real one. The reasons are due to the balancing of an assembly process being conducted unevenly; initial data collected by observation may cause deviations, etc. However, analytical approach enables to solve theoretically many technical problems and predict real output of systems with minor corrections that reflect technical parameters of processes.

The result of study of the car cabins assembly line used for calculation and graphical presentation of the productivity rate is showed in Figure 6.8. The initial data of the assembly line are calculated as follows (Table 6.3):

$t_w = 17.3$ min and $t_a = 4.7$ min are average of the work and auxiliary times at bottleneck station, respectively.

$\dfrac{\theta_c}{z} = \dfrac{(1.5+8)\times 60}{456} = 1.25$ min is the idle time referred to one product due to technical reasons of common machines and mechanisms.

$\dfrac{\theta_{st}}{z} = \dfrac{(4.6+0.5/18)\times 60}{456} = 0.608$ min is the idle time referred to one product due to technical reasons and mechanisms at one station.

$\dfrac{\theta_m + \theta_l}{z} = \dfrac{(4.3+15.9)\times 60}{456} = 2.657$ min is the idle time referred to one product due to managerial and logistics problems.

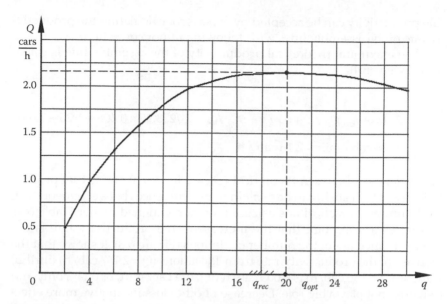

Figure 6.8 Productivity rate of the assembly line versus the number of stations.

After substituting the obtained data into Eq. (6.2) and transforming the equation of theoretical productivity rate of the car cabin assembly line has the following expression

$$Q = \cfrac{1}{\cfrac{185.64 + 47.61}{q} + 0.608q + 1.25 + 2.657} = \cfrac{1}{\cfrac{233.25}{q} + 0.608q + 3.907}$$

Based on Eq. (6.1), the diagram of the productivity rate of the assembly line versus the number of stations is presented (Figure 6.8). The diagram shows that the productivity rate of the assembly serial line is changeable. It increases at the beginning with increments in the number of stations and goes down after reaching some number of stations, which gives maximum productivity.

Optimal number of stations of the assembly line that gives maximum productivity is calculated by Eq. (6.2).

$$q_{opt} = \sqrt{\frac{(t_w + t_{aux})f_c}{\theta_{st} / z}} = \sqrt{\frac{(185.64 + 47.61)1}{0.608}} = 20$$

The optimal number of stations of the assembly line defined theoretically ($q_{opt} = 20$) is close to the actual number of station ($q = 18$) and confirms that company solved the structure of the assembly line by using the best solution. In addition, it proves that theoretical approach of the car assembly line productivity can be accepted by experts for calculating the productivity rate of the assembly lines with following minor corrections.

The maximum theoretical productivity of the assembly line is calculated by Eq. (6.4).

$$Q_{max} = \frac{1}{2\sqrt{(t_w + t_{aux})\theta_{st}/z + \theta_{ct}/z + \theta_{org}/z}} = \frac{1}{2\sqrt{233.25 \times 0.608 + 1.25 + 2.657}}$$

$$= 0.036 \, cab/min = 2.16 \, cab/h$$

The differences between theoretical (Q_{th} = 2.16 cab/h) and actual productivity (Q_a = 2.37 cab/h) can be explained by differences between theoretical work time and practical one calculated that mentioned earlier (Table 6.3). It is necessary to point that the increase of the productivity rate is not even with increase of the number of stations (Figure 6.8). It shows that the increase of the productivity rate from 18 stations (Q = 2.157 cab/h) until 20 stations (Q = 2.16 cab/h) is 0.003 cab/h or 0.024 cab/shift, or 0.14% only and that does not play a big role. Decrease of other losses can give more effectiveness in this case (Figure 6.8). Hence, the optimal number of stations can be reduced to 17–18 recommended stations, which do not give the sensitive drop of productivity rate (Figure 6.8). On the other hand, decreasing the number of stations can reduce the financial expenditures for the assembly line that can be effective. A car company practically defined the optimal number of assembly stations. In addition, computed productivity rate by Eq. (6.1) and actual one is well matched, i.e. mathematical model for productivity rate is validated and can be used for the practitioners at real industrial environment.

On the basis of the obtained results of research (Table 6.3), the assembly line balancing was converted and the auxiliary time of the assembly process reduced. After rearranging, the new maximum cycle time at bottleneck station became $T = t_w + t_a = 14 + 4.5 = 18.5$ min. Hence, the assembly line's new data presented by the following indices:

Limit of the assembly line productivity rate is

$$Q_L^* = \frac{1}{t_w} = \frac{1}{14.0} = 0.0714\frac{car}{min} = 4.28 \, cab/h.$$

Cyclic productivity of the assembling flow line is

$$Q_c^* = \frac{1}{t_w + t_{aux}} = \frac{1}{18.5} = 0.054\frac{car}{min} = 3.24 \, cab/h.$$

The company elaborated solutions to decrease other productivity losses that can improve the output of the assembly line.

The use of mathematical model in solving some of the manufacturing problems has become a common routine. Particularly, engineers solving some of their assembling tasks can use the new mathematical and practical approach of study assembly line wherever suitable. Manufacturers can carry out system analysis of assembly processes to solve the productivity problems. The proposed system analysis of assembly processes is universal and can be applied to any kind of technological process. The presented method include the mathematical model of the productivity rate of an assembly line, which enables to calculate maximum productivity rate, optimal and number of assembly stations recommended, productivity losses as function of technological, technical, managerial and organisational indices of the production processes.

The method showed main productivity losses in the cabin assembly process due to auxiliary motions, reliability of air compressor, conveyor and pneumatic tools and not perfect managerial activity that takes in general 30% of productivity losses. These productivity losses should be reduced by stepped continuous improvement of manufacturing process to increase the output of the car cabin assembly line. Equation of the productivity rate of the assembly line confirmed practical results and showed new approach in identification of the optimal number of assembly stations by criterion of maximum productivity. All findings are useful and can be applied in manufacturing industries to improve the efficiency of production processes. Further improvement of assembly process should be focused on decreasing of the auxiliary and assembly times by constructional, technological and structural approaches. The use of different types of shelf, jigs and mechanisms, which enables to create comfortable conditions that reduce auxiliary time for assembly processes, can solve constructional and technological problems. Structural approach means the assembly serial line can be divided on some sections with embedded buffers. Such solution enables to enhance the productivity of the assembly line in case of low reliability of its mechanisms and machines. The optimal number of the section can be defined analytically by criteria of productivity rate and economic efficiency of the assembly line.

Sustainable Improvement. Final stage in DCAS engineering mathematical analysis method is sustainable improvement. Improvement process is selected by PACE Prioritisation Matrix, which shows in Figure 6.9 based on the data of Tables 6.2 and 6.3. Prioritisation Matrix is the result from the closed group discussion with production manager and engineering departments in a car company. Figure 6.6 clearly demonstrates that the productivity loss is due to many reasons, where the auxiliary time parameter for the assembly line of car cabin is dominated among other losses. This auxiliary time is a component of the cycle time at the bottleneck assembly station and is the result of a not well-balanced assembly process abbreviated as A&B. Analysis of the nature of auxiliary time demonstrates

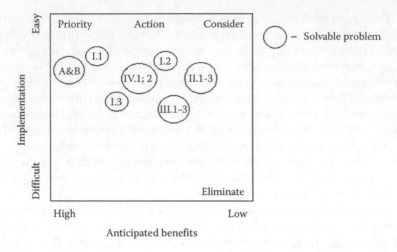

Figure 6.9 PACE Prioritisation Matrix for an assembly line of a car cabin.

that to solve this problem, it is necessary to find the better balancing by rearranging the assembly process. In addition, to decrease the auxiliary time it is possible to reconsider the location of the assembly parts and manual tools at assembly station for comfortable work of employees. All these actions do not represent some complex unsolvable problems and enable decrease in the value of auxiliary time until acceptable level that restricted the physical ability of an assembler.

These actions for decreasing the auxiliary time is represented by the symbol (A&B*) with an asterisk for the PACE Prioritisation Matrix.

The productivity losses of the assembly line of car cabin are represented in Table 6.3, and analysis shows that losses due to technical problems are solved by decreasing the pressure drop for pneumatic tools and by installation of a more powerful compressor. The convenor chassis is not a reliable unit that should be replaced by new one. These actions are represented in the PACE Prioritisation Matrix.

The minor productivity loss due to managerial problems can be decreased as it depends on situational activity of assembly process and considered by a group of managers. Logistic problems are generally managerial problems, and their productivity losses are decreased by improvement of management system that considers the assembly parts supply. It is not a major problem and can be solved by delivery of assembly parts at the prescribed time without additional expenses on assembly process. This problem is a challenge for consideration by the group of managers. The productivity loss due to mistaken assembly is a result of wrong supply of assembly parts. This problem is solved by managers, and in case of similarities of assembly parts, they should be labelled to avoid mistakes in an assembly process. Pipe leaking leads to defected assembly that is solved by improvement of tools.

Detailed analysis of reasons of productivity losses in assembly line did not show some components of production process, which is useless and is a subject of elimination. The actions described earlier enable decrease in the productivity losses of an assembly line and improve the engineering culture of production process. Some of the actions are easy to implement, others takes time or cannot give sensitive improvement for assembly process of a car cabin. All activities for sustainable improvement are represented in PACE Prioritisation Matrix diagram.

Application of DCAS Engineering Mathematical Analysis Method for a car company with the help of mathematical model for productivity rate enabled to compute all productivity rate losses that are measured and analysed with mathematical form. The losses of productivity rate of assembly line can analyse visually, clearly, systematically and effectively which is much better than just simply analysis of statistical data. The mathematical model of productivity rate is validated and proved that is useful and more accurate than other previously used methods. This mathematical model of productivity rate and DCAS Engineering Mathematical Analysis Method can apply to different fields of industrial engineering.

6.3   *Productivity attributes and management analysis for manufacturing systems*

Manufacturing processes in industry are complex and involve activities from different departments. Many indices are evaluated in the manufacturing process, but the two of note are the productivity of manufacturing systems and the quality of products. The work of many different departments is concerted and coordinated to carry out the main indicators of the production system. This complex coordination has causative, consequential and effective links that should be formulated using mathematical models. Such models are useful in evaluating the efficiency of production system components, and numerically, they demonstrate the level of their influence on factory outputs.

The amount of production time that is spent on manufacturing processes should be used effectively and should be evaluated by attributes that can give a clear picture of how time is used. Actual production processes depend on many factors, including technology, machine and systems design, machine exploitation and service systems, the reliability and maintenance of machinery, managerial and organisational aspects in the facilities. For example, concerning the latter, production processes based on the concerted activity of different industrial departments and divisions enable the use of production time effectively and efficiently.

However, real-world industrial environments demonstrate that production time is not used fully during manufacturing processes. Conducted analysis of actual production times demonstrated that typical components of time are permanent in manufacturing processes presented in mathematical form.

Production processes are evaluated mainly by the attribute known as productivity rate, for which the dimensions depend largely on the type of industry and the technologies used. Production processes are complex systems and depend on many factors mentioned earlier. Different complexities in design and technology of manufacturing systems demonstrate different productivity rates. Analysis of reasons of these differences enables to evaluate the weight in efficiency of all components of technology and design in mathematical model of productivity rate. Such an approach makes it possible to consider separately the properties of manufacturing system in reality.

Analysis of the production time of an industrial machine or complex manufacturing system demonstrates the following: the tendency in industries is to create machines and systems with complex designs that can lead to an increase in cyclic productivity with respect to segmentation and duplication processes, which ultimately decreases the availability, which represents the design complexity of machinery. The values of cyclic productivity and design complexity factor are changing in opposite directions. Productivity rate is increased with increase in the design complexity of machinery. Design complexity of manufacturing systems leads to a decrease in the value of availability and leads to a decrease in the productivity rate. These factors demonstrate the reasonable links and their interrelations in engineering problems that should be mathematically described and evaluated. Such approach enables to predict the output and efficiency of production system.

Productivity rate depends on many factors and each of them describes one side of the production process. Positive and negative properties of manufacturing system should be in balance to give effective result. The production system has a defined limit in terms of its productivity rate. Hence, the structure of such a production system can be optimised by economic criterions and defining a maximum productivity rate.

Evaluation of the manufacturing processes should be started from analysis of the time spent on production process. First step is the analysis of the equation for productivity rate per cycle time (Eq. 2.3) that demonstrates the following: decrease in cyclic machining time t_m leads to an increase in productivity rate of industrial machines. The principle of intensification of machining processes has several limitations due to the inherent properties of technological processes and quality of machining products. Practically, decrease in machining times is achieved by a balance of technological process that leads to constructional complexity of automated industrial machines and systems. This method leads to a decrease in machining cycle times and increase in the design complexity of industrial machines expressed by their availability. This conclusion can be demonstrated on some typical or routine examples.

The productivity rate per cycle time for a single machine is represented by the following equation [16,21]:

$$Q = \frac{1}{(t_m + t_a)} \times \frac{1}{1 + m_r(\lambda_s + \lambda_{cs})} \tag{6.5}$$

where $Q_c = 1/(t_m + t_a)$ is the cyclic productivity of the single machine and $A = 1/[1 + m_r(\lambda_s + \lambda_{cs})]$ is the availability of the single machine that expresses its design complexity [17].

The productivity rate per cycle time for an automated line of serial structure is represented by the following equation:

$$Q = \frac{1}{\left(\dfrac{t_{mo}}{q} f_c + t_a\right)} \times \frac{1}{1 + m_r \left(\displaystyle\sum_{i=1}^{q} \lambda_{s.i} + \lambda_{tr} + \lambda_{cs}\right)} \tag{6.6}$$

where $Q_c = 1/\left(t_{mo} f_c / q + t_a\right)$ is the balanced cyclic productivity of the automated line of serial structure and $A = 1/\left[1 + m_r\left(\displaystyle\sum_{i=1}^{q} \lambda_{s.i} + \lambda_{tr} + \lambda_{cs}\right)\right]$ is the availability of the automated line that expresses its design complexity.

The productivity rate per cycle time for an automated line of parallel–serial structure segmented on sections with buffers is represented by the following equation:

$$Q = \frac{p}{\dfrac{t_{mo}}{q} f_c + t_a} \times \frac{1}{\left[1 + m_r\left(\dfrac{pf_s \displaystyle\sum_{i=1}^{q} \lambda_{s.i}}{n} + \Delta\lambda_{f(i-k)} + \cdots + \Delta\lambda_{f(i-2)} + \Delta\lambda_{f(i-1)} + \Delta\lambda_{b(i+1)} + \Delta\lambda_{b(i+2)} + \cdots + \Delta\lambda_{b(i+n)} + \lambda_{bf} + \lambda_c + \lambda_{tr}\right)\right]}$$

$$\tag{6.7}$$

where two components are balanced cyclic productivity and availability.

The mathematical models for productivity rates of other industrial systems of complex structures contain two similar components (Eqs. 6.5–6.7). Analysis of their mathematical models demonstrates the following regularities and reasonable links in evolution of manufacturing machines and systems. Technology and machine design is a system that cannot be separated. The change in technology leads to change in machine design. The expressions of the cyclic productivity Q_c contain symbols of technology t_{mo} and symbols of design of machine q, p and t_a which is the auxiliary time. The latter one is linked with the dynamical characteristics of the mechanical components of the manufacturing machines. The availability A expresses design principles and reliability of the machine or system. Both components relate to the constructional complexity of machine or system.

Similar analysis can be conducted for the other types of manufacturing systems and automated lines with different structures. This statement confirms that increase in the complexity of a machine's design leads to an increase in the value of its cyclic productivity and a decrease in the value of its availability (Figure 6.10). All factors are interrelated, and finding the optimal ratio between them is solved by economic indices.

The machine productivity Q represents the product of cyclic productivity and machine availability. An increase in machine constructional and structural complexity has the right links with the segmentation, balancing and intensification of the technological process. Latter one is limited due to physical principles of technology and quality of machining processes. Practically, industrial systems have tolerances on auxiliary time and availability, which are evaluated on appropriateness by economic indices. Manufacturers prefer to reduce variations in factors' tolerance, but any attempt to improve the values of the auxiliary time and availability leads to changes in other parameters of productivity rate. This is the reason that the availability for automatic machines and automated lines should not be less than 0.85. This limit partially is based on a preference for short times for auxiliary motions. Decreasing this time leads to an increase in in the dynamic loads on machine components and reduces machine availability and the quality of the product manufactured. Additionally, increase in

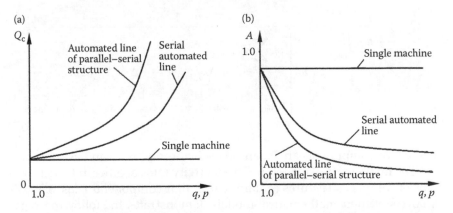

Figure 6.10 The change in technological productivity (a), design factor and availability (b) versus change in the complexity of a machine and system design.

machine reliability leads to an increasing in the cost of machines. Structural complexity of the manufacturing systems leads to an increase in cyclic productivities and to decrease in the values of availability (Figure 6.10a,b). All these components demonstrate interrelations of the implemented technology, design of manufacturing system and its reliability. Optimal correlation of these parameters by criterion of the maximal productivity rate and minimal cost is available by mathematical modelling.

The management and organisational factors of the production system play a big role in its productivity rate. However, the management factor is not a direct part of the manufacturing process but is rather linked to the availability and efficiency of a production system. As far as the technological and technical parameters are the basis of manufacturing machines and systems design, the management and organisational factor is omitted from analysis in mathematical models for productivity rates. The management and organisational factor is considered separately in a structure of the control of production processes.

Working examples

Productivity rates of manufacturing systems and the factors that lead to decreasing production system outputs are shown in Eqs. (6.5–6.7). An analysis of the productivity rates and factors for a typical manufacturing machine and systems is given later. An engineering problem is represented, and the productivity rates and factors calculated for the given data of the manufacturing machine shown in Table 6.4; the technical and timing data of the manufacturing machine refer to one product. These results are obtained by substituting the initial data (Table 6.4) into components of Eqs. (6.5–6.7). The calculated results

Table 6.4 Typical technical data of manufacturing machines and systems

Technical data		Single machine	Serial automated line	Parallel–serial automated line
Machining time, t_{mo}		5.0 min		
Auxiliary time, t_a		0.2 min		
Number of workstations	Serial, q	1	10	10
	Parallel p_s			2
	Correction factor, f_c	1.0	1.2	
Failure rate, $\lambda \times 10^{-4}$ (per min)	Workstation,	40		
	Control system	5		
	Transport system	2		
Mean time to repair, m_r (min)		3		

represented in Table 6.4 facilitate the evaluation of the attributes of the productivity rates for the manufacturing machine.

The following example demonstrates that an increase in the design complexity of the manufacturing machine or systems leads to a decrease in the value of availability [18]. Table 6.4 represents the technical data of typical industrial machines and systems.

Table 6.5 shows the values for the cyclic productivity and availability (reliability index) of industrial machines and systems with complex designs.

A graphical representation of the increase in cyclic productivity and decreasing of availability for the industrial manufacturing machine and systems is depicted in Figure 6.11.

Figure 6.11 clearly depicts the dependency of decrease in availability values and the increase in cyclic productivity of manufacturing machine and systems. These variables should be in balance that gives maximum productivity rate. Optimisation of structure for the manufacturing systems with different designs by criterion of maximal productivity rate is conducted by mathematical methods represented in previous chapters.

6.4 Manufacturing system availability and number of technicians

Availability for complex manufacturing systems have right links to their productivity rate and the number of technicians that serve these systems. Manufacturing systems are evaluated by the value of reliability, and practically, technicians support the workability of the machinery. The design of a perfect manufacturing system with high reliability is theoretically possible, but this system will have very high cost that is practically unacceptable. Real manufacturing systems are serviced by technicians that support the system in workable conditions [20]. Manufacturing system can be represented by different structures and layouts. It can be automated line or group of single machines, etc. The downtimes of the single machines or workstations that work independently from other machines do not coincide in time and in duration. Technicians can serve one machine, but at the same time, other machines can have downtimes that lead to the idle time of machinery. It is known that technicians who conduct operations of machine inspection service and eliminate random failures of manufacturing machines, and support production systems workability. It means technicians are not conducting all production time in repair. How many technicians should be in service of the manufacturing system is a typical problem for industries. To find the solution for this problem, the following method is proposed.

Modern industrial machines and systems with complex designs require reliable mathematical models to calculate the manufacturing

Table 6.5 Cyclic productivity, design factor and availability of manufacturing machine and systems with complex designs

Title	Single machine	Serial automated line	Parallel–serial automated line
Equation			
		Cyclic productivity (product/min)	
	$Q_c = 1/(t_m + t_a) = 1/(5.0 + 0.2)$ $= 0.192$	$Q_c = 1/[(t_{mo}/q)f_c + t_a] = 1/[(5.0/14) \times 1.2 + 0.2] = 1.591$	$Q_c = p/[(t_{mo}/q)f_c + t_a] =$ $2/[(5.0/14)1.2 + 0.2] = 3.182$
		Availability	
	$A = \dfrac{1}{1 + m_r(\lambda_s + \lambda_c)}$ $= \dfrac{1}{1 + 3.0(40 + 5)10^{-4}}$ $= 0.988$	$A = \dfrac{1}{1 + m_r(q\lambda_s + \lambda_c + \lambda_{tr})}$ $= \dfrac{1}{1 + 3.0(10 \times 40 + 5.0 + 2.0)10^{-4}}$ $= 0.89$	$A = \dfrac{1}{1 + m_r(p_s q\lambda_s + \lambda_{tr} + \lambda_c)}$ $= \dfrac{1}{1 + 3.0(2 \times 10 \times 40 + 2 + 5)10^{-4}}$ $= 0.805$

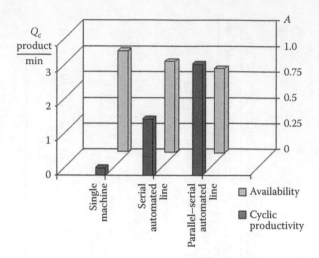

Figure 6.11 Cyclic productivity and availability of manufacturing machine and systems.

system's productivity rate. The productivity rate for any manufacturing machines and systems depends on the technological processes, the reliability of machines and mechanisms and how managerial and organisational problems are dealt with. Researchers proposed several methods for defining the number of technicians for inspection and repair the machinery in workable conditions. These methods are considering the repair random failures of machines and planned preventive repair systems. However, practically, these methods are corrected and managers look for reliable mathematical solution for calculation of the number of technicians. Solving the problem of optimal number of technicians for the manufacturing systems that support high efficiency of machinery at regular production time is a crucial issue.

Methodology for calculation of the manufacturing system reliability, maintainability and replacement is represented by several key publications that consider all aspects of manufacturing processes. Intensification of industrial processes based on the manufacturing systems with complex design enhanced their reliability problems. Today, efficiency of expensive manufacturing systems like automated lines with complex design depends on the number of technicians, which provide support system in workable condition. The number of technicians for the service of manufacturing system should be defined by mathematical modelling of production processes. Decrease in the reliability of automated manufacturing systems leads to decrease in the productivity rate and increase in the number of technicians.

Reliability problem of machinery and the number of technicians in servicing industry are not new, and numerous publications and different models enable describing analytically manufacturing problems. The probability and reliability theories provide standard attributes and modes of calculation that can be used for the mathematical modelling of the number of technicians in servicing of machines. However, most of the known publications describe system of maintenance, reliability and replacement in engineering at preventive and planned time. There are publications which describe the reliability of manufacturing systems and service system with technicians using a probabilistic approach that allow for the complicated or simplified calculation of the average searching parameters.

The tendency of engineering progress is represented in the form of employment industrial machines with complex and expensive designs, whose workability should be supported by accurate service system. Mathematical models of industrial machines' availability are derived according to the level of consideration of the manufacturing processes. The method for calculation of the number of technicians is based on availability and probabilistic failure rates of a manufacturing system. The number of technicians for servicing, inspection and repair is defined for random failures of machinery and do not consider aspects of maintenance, repair and replacement of machines at prescribed, preventive and planned overhaul repair time. The concept of maintenance considers the machine when it is out of work, i.e. the planned repair and service of machines are conducted for stopped overhaul processes that do not include the random downtimes of machines.

The reliability theory represents the following indices: machine failure rate $\lambda = 1/m_w$; mean time to work m_w; mean time to repair m_r and availability A. These attributes of reliability are used in equations for the productivity rate of manufacturing systems and can be used for calculation of the number of machines in service by one technician.

6.4.1 Methodology

Availability for complex manufacturing systems has right links with their productivity rate and the number of technicians who serve these systems. Manufacturing systems are evaluated by the level of reliability and, practically, technicians support the workability of the machinery. Theoretically, it is possible to design perfect manufacturing systems with high reliability, but this system will have very high cost that is practically unacceptable. Real manufacturing systems are serviced by technicians to support the system in workable conditions. Manufacturing system can be presented by different structures and layouts. It can be automated line or group of single machines, etc. The downtimes of the single machines or stations that work independently from other machines do not coincide in time and duration. Technicians can service one machine, but at the same time, other machines can have downtimes that lead to an idle time of machinery.

Usually technician's duty is to conduct repairs, support and serve in workable condition machines. It means technicians are not conducting all production time in repayment. How many technicians should be in service for a manufacturing system is a typical problem for industries. To find the solution for this problem, the following method is proposed.

The attribute availability of machines from reliability theory can be used for calculating the number of technicians that is necessary to support a machine or system in workable condition. The equation of machine's availability is expressed by Eq. (1.12), whose transformation gives the failure rate of a machine presented by the following:

$$\lambda = \left(\frac{1}{A} - 1 \right) \Big/ m_r \qquad (6.8)$$

Equation (1.12) represents dimensionless time or probability of machine's work. The probability of machine's downtime due to reliability reasons is represented by Eq. (2.11):

$$I_{i.A} = \left(1 - \frac{1}{1 + m_r \lambda} \right) \qquad (6.9)$$

The number of random repairs of machines per hour by the technician can be calculated by the following formula:

$$n = \frac{1\,\mathrm{h}}{m_r} \qquad (6.10)$$

where n is the number of random repairs that is equal to the number of failures.

The maximal number of machines serviced by one technician is defined by the following formula:

$$q_{max} = \frac{n}{\lambda} = \frac{1}{m_r \lambda} \text{ or } m_r \lambda = \frac{1}{q_{max}} \qquad (6.11)$$

This number of machines in service is calculated when failures of machines do not coincide at one time. Practically, machines' downtimes are coinciding randomly and can be calculated by rules and regulations of probability theory. Probability that all machines of independent work will have downtime at one time is calculated as the product of probabilities of failures of all machines according to the rules of probability theory. This probability is calculated by the following equation:

$$P = P_1 \times P_2 \times P_3 \times \cdots \times P_q \qquad (6.12)$$

In engineering, the probability of density distribution of failures is represented by exponential function. Probability that machine will work at some defined time without failures is calculated by the following equation:

$$R(t) = e^{-\lambda t} \tag{6.13}$$

where t is the defined time, and all other parameters are specified earlier.

Probability that one machine will have failure during one-hour work is represented by the following equation:

$$P_1(t) = 1 - R_1(t) = 1 - e^{-\lambda_1 t} \tag{6.14}$$

Probability that q machines with different reliability have failures at one time is represented by the following equation:

$$P_q(t) = \left[1 - R_1(t)\right] \times \left[1 - R_2(t)\right] \times \left[1 - R_3(t)\right] \times \cdots \times \left[1 - R_q(t)\right] \tag{6.15}$$

Substituting Eq. (6.13) into Eq. (6.15) yields the following equation:

$$P_q(t) = (1 - e^{-\lambda_1 t}) \times (1 - e^{-\lambda_2 t}) \times (1 - e^{-\lambda_3 t}) \times \cdots \times (1 - e^{-\lambda_q t}) \tag{6.16}$$

Probability that n machines with same reliability will have failures at one time is represented by the following equation:

$$P_q(t) = \left[1 - R(t)\right]^q = [1 - e^{\lambda t}]^q \tag{6.17}$$

Practically, the number of machines that failed at one time is going on randomly. However, this coinciding of n failures leads to an increase in the downtime of machines that can be represented via availability and have probabilistic nature. Hence, the probability of idle time for machines that have different reliability level can be represented by the following equation:

$$\sum_{i=1}^{q} I_{i.A} = 1 - \sum_{i=1}^{q} \frac{1}{1 + m_r \lambda_i} \tag{6.18}$$

The probability of idle time for machines that have same reliability level can be represented by the following equation:

$$I_{i.A} = 1 - \left(\frac{1}{1 + m_r \lambda}\right)^q \tag{6.19}$$

where q is the number of machines failed at one time, and other parameters are as specified earlier.

Equations (6.16) and (6.17) are dimensionless and contain the number of machines n that randomly failed at one time. Increase in the number of machines failed at one time leads to an increase in the downtime of machines with decrease in their probability of implementation. Increase in the number of machines failed at one time leads to a decrease in the probability of downtime for machines. The graphical presentation of these two functions versus the number of machines failed at one time is represented in Figure 6.12.

The diagram depicts that the downtime I of manufacturing system with n machines is growing because of simultaneous failures of several machines. However, the probability P of coincided failures of machines is decreasing. In such condition, the optimal number of machines serviced by one technician is a complex problem but can be solved by two approaches. The first approach is based on average probability of coincided failures of machines that accepted probability theory. The second approach is defined by economical method based on the following solution. Increase in the downtimes of machines leads to a decrease in the productivity rate of manufacturing systems and decrease in income. Increase in the number of technicians to support workable condition of the manufacturing system leads to an increase in the cost of machine service. These two economic aspects should give optimal number of machines in service by one technician that can minimise the productivity losses and hence optimise effectiveness of manufacturing system.

Practically, the complex multi-station manufacturing system services by the technician crew which collaborate work enables following minimisation of downtimes. One technician cannot decrease downtimes of coincided failures of several machines due to engagement with one machine while others are not in workable conductions. In this case, the work of

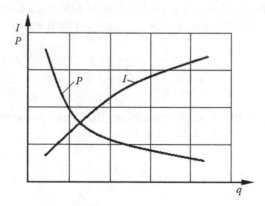

Figure 6.12 Probability of downtime and probability of the number of machines failed at one time versus the number of machines.

technician crew for servicing complex manufacturing system should be arranged by management system to decrease the minimum downtime by cooperative work.

The approach with average probability of coincided failures of machines with equal failure rates enables to give the following solution. The number of machines in service by one technician presented by Eq. (6.11) for the failed one machine with probability is expressed by Eq. (6.12). The probability of coincided failures of n machines is represented by Eq. (6.15). The number of machines that failed at one time is going on randomly, and for the calculation of the technical parameters for a manufacturing system is used the average probability of coincided failures of machines. Such method is normal practice in area of probabilistic analyses and calculations. Hence, the number of machines in service by one technician should be calculated by the following corrected equation:

$$q_{sr} = \frac{1}{m_r \lambda} \times f_{cor} = \frac{1}{m_r \lambda} \times \frac{\displaystyle\sum_{i=1}^{q} P_q(t)}{q_{max} P_1(t)} = \frac{1}{m_r \lambda} \times \frac{\displaystyle\sum_{i=1}^{q} (1 - e^{\lambda_i t})^q}{q_{max}(1 - e^{\lambda_1 t})} \qquad (6.20)$$

where $f_{cor} = \dfrac{\displaystyle\sum_{i=1}^{q}(1 - e^{\lambda_i t})^q}{q_{max}(1 - e^{\lambda_1 t})}$ is the correction factor, $P_{av}(t) = \dfrac{\displaystyle\sum_{i=1}^{q}(1 - e^{\lambda_i t})^q}{q_{max}}$ is the average probability of coincided failures of machines, and other parameters are as specified earlier.

Equation (6.20) presents the corrected number of machines in service by one technician with random coincided failures of group of machines. Correction factor reduces the number of machines in service, because one technician cannot serve other failed machines at one time. The number of machines with different reliability in service by one technician should be calculated by the following equation:

$$q_{sr} = \frac{1}{m_r \lambda_{av}} \times f_{cor}$$

$$= \frac{1}{m_r \lambda_{av}} \times \frac{\displaystyle\sum_{i=1}^{q}(1 - e^{-\lambda_1 t}) \times (1 - e^{-\lambda_2 t}) \times (1 - e^{-\lambda_3 t}) \times \cdots \times (1 - e^{-\lambda_q t})]}{q_{max}(1 - e^{-\lambda_1 t})} \qquad (6.21)$$

where λ_{av} is the average failure rates for group of machines, and all parameters are as specified earlier.

Practically, the mean time to repair random failures in industries presents by the short time and in average is $m_r = 2,..., 3\,\text{min}$. It means that one technician can conduct in average $n = 20,...,30$ repairs per hour. Substituting Eq. (6.20) into Eq. (6.7) and transformation yields the following equation:

$$A = \frac{1}{1 + m_r \lambda} = \frac{1}{1 + \dfrac{1}{q_{sr}}} = \frac{1}{1 + \dfrac{1}{q_{max} f_{cor}}} = \frac{1}{1 + \dfrac{1}{q_{max}} \times \dfrac{q_{max}(1 - e^{\lambda_1 t})}{\displaystyle\sum_{i=1}^{q}(1 - e^{\lambda_i t})^q}} \quad (6.22)$$

where $q_{sr} = q_{max} f_{cor}$, and all parameters are as specified earlier.

Transforming Eq. (6.22) and expressing in term of the corrected number of machines q_{sr} in service by one technician give the following equation:

$$q_{sr} = \frac{\displaystyle\sum_{i=1}^{q}(1 - e^{\lambda_i t})^q}{[(1 / A) - 1][q_{max}(1 - e^{\lambda_1 t})]} \quad (6.23)$$

Equation (6.23) can be represented graphically in Figure 6.13 that demonstrates the change of the number of machines in service by one technician with change of availability of a machine.

Figure 6.13 depicts the exponential increase in the number of machines in service by one technician with increase in the availability of machines. Reliable machines need less number of technicians for service, inspection and repair of random failures.

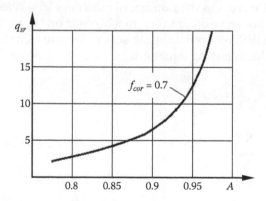

Figure 6.13 Change of the number of machines in service by one technician versus availability of a machine.

A working example 1

A manufacturing system of $q = 20$ machines with equal reliability and independent work are presented with the following attributes: the failure rate $\lambda = 2.4$ failures/hour and $m_r = 2.5$ min is the repair time of random failures. The failure rate of the machine is described by the exponential probability of density function. Calculate the availability and number of technicians for regular service, inspection and repair for random failures of machines.

Solution: The availability of the machine is as follows:

$$A = \frac{1}{1+m_r\lambda} = \frac{1}{1+2.5\times2.4/60} = 0.9090$$

The maximal number of machines in service per hour by one technician is defined by the following formula:

$$q_{max} = \frac{60}{m_r\lambda} = \frac{60}{2.5\times2.4} = 10 \text{ machine/h.}$$

This number of machines is maximal for service by one technician if machines do not have simultaneous failures, and hence no downtimes of other machines. However, practically machines failures are not in turn and downtimes of machines are increased with increased number of simultaneous failures.

Probability that machine will work for one hour without failures is calculated by the following equation:

$$R(t) = e^{-\lambda t} = e^{-[(2.4)/1]\times1} = 0.092$$

This probability is quite low and expected that probability of failure is high. Probability that one machine will have failure during one-hour work is calculated by the following equation:

$$P_1(t) = 1 - R_1(t) = 1 - 0.092 = 0.9078$$

This probability is quite high and expected that downtime of machines is high.

Probability that 10 machines with same reliability will have failures at one time is represented by the following equation:

$$P_n(t) = 0.9078^{10} = 0.380$$

The probability of simultaneous failed 10 machines is high and should be considered for the following calculations.

The average probability of failures for 10 machines is as follows:

$$P_{av}(t) = \frac{\sum_{i=1}^{q}(1-e^{\lambda t})^q}{q} = \frac{0.9078 + 0.9078^2 + 0.9078^3 + \cdots + 0.9078^{10}}{10} = 0.61039$$

Correction factor is as follows

$$f_{cor} = \frac{\sum_{i=1}^{q} (1-e^{\lambda_i t})^q}{q(1-e^{\lambda_1 t})} = \frac{0.61039}{0.9078} = 0.672$$

The corrected number of machines in service by one technician is as follows:

$$q_{sr} = \frac{n}{\lambda} \times f_{cor} = 10 \times 0.672 \approx 7 \text{ machines.}$$

The total number of technicians for service of 20 machines is as follows:

$$q_{sr.tot} = q / q_{sr} = 20 / 6.72 = 2.97 \approx 3 \text{ technicians.}$$

The calculated number of machines in service by one technician is matched with the practical number of machines accepted in industries for the availability of machines $A = 0.9090$. Such number of machines in service by one technician enables him to conduct current service and inspection of machines accept repair of random failures. This optimised solution leads to minimisation of the machines' random downtimes in an industry.

A working example 2

A manufacturing system of eight machines with different reliability attributes and independent work are represented in Table 6.6. The failure rate of the machine is described by the exponential probability of density function. The number of technicians for regular service, inspection and repair for random failures of machines are calculated.

Solution: The number of failures that the technician processes per one hour is calculated by the following expression:

$$n = \frac{1 \text{ h}}{m_r} = \frac{60 \text{ min}}{3.0 \text{ min}} = 20.0 \text{ failure/h.}$$

Table 6.6 Manufacturing system parameters

Title		Number of machines, q	Failure rates per hour, λ		Availability, A	
Group of machines	1	2	2.0	Average, $\lambda_{av} = 2.4$	0.909	Average, $A_{av} = 0.892$
			1.5			$A_{av} = 0.892$
	2	1	2.5		0.888	
	3	3	2.8		0.877	
	4	2	2.3		0.896	
Mean time to repair random failure, $m_r = 3.0$ min						

The maximal number of machines in service per hour by one technician is defined by the following formula:

$$q_{max} = \frac{60}{m_r \lambda_{av}} = \frac{60}{3.0 \times 2.4} = 8.333 \text{ machine/h.}$$

This number of machines is maximal for service by one technician if machines do not have simultaneous failures, hence no downtimes of machines. However, practically, machine's failures are not in turn and downtimes of machines are increased with increased number of simultaneous failures.

Probability that machines in one group will work one hour without failures is calculated by the following equation:

$$\text{Group 1, } R(t) = e^{-\lambda t} = e^{-2.0} = 0.1371$$

$$\text{Group 2, } R(t) = e^{-\lambda t} = e^{-2.5} = 0.0834$$

$$\text{Group 3, } R(t) = e^{-\lambda t} = e^{-2.8} = 0.0619$$

$$\text{Group 4, } R(t) = e^{-\lambda t} = e^{-2.3} = 0.1018$$

Probability that one machine will have failure during one-hour work is calculated by the following equation:

$$\text{Group 1, } P_1(t) = 1 - R_1(t) = 1 - 0.1371 = 0.8629$$

$$\text{Group 2, } P_2(t) = 1 - R_2(t) = 1 - 0.0834 = 0.9166$$

$$\text{Group 3, } P_3(t) = 1 - R_3(t) = 1 - 0.0619 = 0.9381$$

$$\text{Group 4, } P_4(t) = 1 - R_4(t) = 1 - 0.1018 = 0.8982$$

The following represent the probability that eight machines with different reliability rate will have failures at one time:

$$P_{8.av}(t) = \left(\sum_{i=1}^{8} P_i(t) \right) \Big/ 8 = (0.8629 + 0.8629^2 + 0.8629^2 \times 0.9166$$

$$+ 0.8629^2 \times 0.9166 \times 0.9381 + 0.8629^2 \times 0.9166 \times 0.9381^2$$

$$+ 0.8629^2 \times 0.9166 \times 0.9381^3 + 0.8629^2 \times 0.9166 \times 0.9381^3 \times 0.8982$$

$$+ 0.8629^2 \times 0.9166 \times 0.9381^3 \times 0.8982^2) / 8 = 0.6139$$

The probability that one machine with the average failure rate will work one hour without failures is as follows:

$$P_1(t) = 1 - e^{-\lambda_{av} t} = 1 - e^{-(2.4 \times 1)/1} = 0.9078$$

Correction factor is as follows:

$$f_{cor} = \frac{\sum_{i=1}^{q}(1-e^{\lambda_i t})^q}{q(1-e^{\lambda_{av}t})} = \frac{0.6139}{0.9078} = 0.6762$$

The corrected number of machines in service by one technician is as follows:

$$q_{sr} = \frac{n}{\lambda} \times f_{cor} = 8.333 \times 0.6762 = 5.6354 \text{ machines.}$$

The total number of technicians for service of eight machines is as follows:

$$q_{sr.tot} = q / q_{sr} = 8 / 5.6354 = 1.419 \text{ technicians.}$$

The calculated number of machines in service by one technician and the number of technicians are not integer numbers. Average availability is not high $A = 0.892$, and this is the reason of low number of machines in service by one technician. These machines and others with independent work and different reliability should be serviced with crew and the manager should coordinate the work to minimise the downtime of manufacturing system.

Real manufacturing system is serviced by technicians to support the system in workable conditions. Technicians are conducting service, preventive observation and repair of random machine failures. The random downtimes of the single machine that work independently on other machines can coincide in time and in duration with downtimes of other machines. Technician who serves one machine cannot serve other machines that have downtimes at the same time. This circumstances lead to an idle time of machinery. Methodology for calculation of the number of machines in service by one technician is based on the availability of a machine with failure rate and mean time to repair, and on the probability of the coinciding of downtimes of machines. Results of proposed method are matching with practical normative data for servicing the number of machines by one technician. Derived mathematical model and method for calculating the number of machines in service by one technician can be accepted for practical use.

Bibliography

1. Badiru, A.B., and Omitaomu, O.A. 2011. *Handbook of Industrial Engineering Equations, Formulas, and Calculations.* Taylor & Francis. New York.
2. Ben-Daya, M., Duffuaa, S., Raouf, A. 2000. *Maintenance, Modeling, and Optimization.* Kluwer Academic Publishers. New York.

3. Benhabib, B. 2005. *Manufacturing: Design, Production, Automation, and Integration.* 1st edn. Taylor & Francis. New York.
4. Birolini, A. 2007. *Reliability Engineering: Theory and Practice.* 5th edn. Springer. New York.
5. Boothroyd, G. 2005. *Assembly Automation and Product Design.* 2nd edn. Taylor & Francis. New York.
6. Chryssolouris, G. 2006. *Manufacturing Systems: Theory and Practice.* 2nd edn. Springer. New York.
7. Ebeling, C.E. 2010. *An Introduction to Reliability and Maintainability Engineering.* Waveland Press Incorporated. Long Grove, IL.
8. *Groover, M.P.* 2013. *Fundamentals of Modern Manufacturing: Materials, Processes, and Systems.* 5th edn. (Lehigh University). John Wiley & Sons.
9. Games, G. 2010. *Modern Engineering Mathematics.* Prentice Hall. London.
10. Jardine, A.K.S., and Tsang, A.H.C. 2013. *Maintenance, Replacement, and Reliability: Theory and Applications.* 2nd edn. CRC Press. New York.
11. Kalpakjian, S., and Schmid, S.R. 2013. *Manufacturing Engineering & Technology.* 7th edn. Pearson. Cambridge.
12. O'Connor, P.D.T., and Kleyner, A. 2012. *Practical Reliability Engineering.* 5th edn. John Wiley & Sons. West Sussex.
13. Ortiz, C. 2006. *Kaizen Assembly: Designing, Constructing, and Managing a Lean Assembly Line.* Taylor & Francis. New York.
14. Salvendy, G. 2007. *Handbook of Industrial Engineering: Technology and Operation Management.* 3rd edn. John Wiley & Sons. Printed online.
15. Shaumian, G.A. 1973. *Complex Automation of Production Processes.* Mashinostroenie. Moscow.
16. Sin, T.C. et al. 2015. Engineering mathematical analysis method for productivity rate in linear arrangement serial structure automated flow assembly line. *Mathematical Problems in Engineering.* Article ID 592061.
17. Usubamatov, R., Ismail, K.A., and Shah, J.M. 2012. Mathematical models for productivity and availability of automated lines. *International Journal of Advanced Manufacturing Technology.* DOI 10.1007/s00170-012-4305-y.
18. Usubamatov, R., Sin, T.C., and Ahmad, R. 2016. Mathematical models for productivity of automated lines with different failure rates for stations and mechanisms. *The International Journal of Advanced Manufacturing Technology.* DOI 10.1007/s00170-015-7005-6.
19. Usubamatov, R., Riza, A.R., and Murad, N.M. 2013. A Method for assessing productivity in buffered assembly processes. *Journal of Manufacturing Technology Management.* 24(1). pp. 123–139.
20. Usubamatov, R., and Bhuvenesh, R. 2015. Method of assessing the number of technicians in service of manufacturing system. *Journal of Manufacturing Science and Production.* DOI 10.1515/jmsp-2015-0005.
21. Volchkevich, L.I. 2005. *Automation of Production Processes.* Mashinostroenie. Moscow.
22. Wallace, E., Blischke, R., and Murthy, D.N.P. 2004. *Case Studies in Reliability and Maintenance.* John Wiley & Sons. New York.

chapter seven

Conclusion

Production systems are complex, for which basic manufacturing and assembly workshops are equipped by facilities and machineries represented by different machine tools and manufacturing lines that fabricate products. Manufacturing processes focused on high output depend on technology, design and reliability of machinery and the system of exploitation used. Real industrial practice demonstrates that several factors have an impact on the production system output. Apart from these factors, the management and organisational factor plays a significant role in the production system. All stages of the production system are solving problems by economic methods, for which the main component is productivity rate. The productivity rate of a manufacturing system is a very weighty economic index that should be, as much as possible, accurately predicted and evaluated using analytical methods and validated by practical data.

The mathematical models for the productivity rate of a manufacturing machines or systems allow for the analysis and calculation of the values for the technical and technological factors and for the economic analysis of the components of production systems. The production machinery can use manufacturing systems of different structures that depend on the type of production system. In industrial areas, many production systems may be viewed as assemblies of many interacting manufacturing automated lines. In even more complex cases, serial and parallel structures of manufacturing lines are intermixed. Practically, the automated lines are often arranged in mechanical and logical serial, parallel and mixed structures according to the technological process for manufacturing products. Figure 7.1 demonstrates a typical complex manufacturing system combined by linear, parallel and parallel–linear structures of automated lines with buffers. This complex manufacturing system contains eight subsystems with buffers: subsystem 1 represents the automated line with two parallel and one serial structure; subsystem 2 represents the automated line with one serial structure; subsystem 3 represents two automated lines with independent serial structures; subsystem 4 is the same as subsystem 1; subsystem 5 represents the automated line of serial structure segmented on three sections with buffers; subsystems 6 and 7 represent the automated lines of parallel structure with an embedded buffer and subsystem 8 is the same as subsystem 2. This manufacturing system and

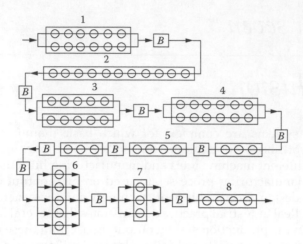

Figure 7.1 Typical complex structure of manufacturing systems with automated lines.

others should be computed on productivity rates that respond on defined economic indices of production system.

Such combinations of manufacturing automated lines serial and parallel structures with buffers cannot be represented by a single mathematical model for productivity rate. If the manufacturing system consists of a combination of series and parallel manufacturing lines, engineers often apply very convoluted block equations, in which the application is problematic.

However, practitioners should solve the problem of computing the productivity rates for combined manufacturing systems with complex structures. From previous presentations in Chapters 2–5 of equations, the productivity rates for serial and parallel manufacturing lines are easily calculated. To calculate the productivity rate for combined complex manufacturing systems represented in Figure 7.1, it is necessary to use the method of simplification of combined structures. The main idea of such method is bringing the complex combined structures of the automated line to a simple structure. Practically, a large number of automated lines with different structures should be segmented on serial subsystems and each of them is computed by own equation of the productivity rates. These eight subsystems of parallel–serial structures with equal productivity rate are represented as the consecutive chain of automated lines with serial structures and segmented on buffered sections. Each automated serial line is described by the technological, technical and reliability parameters. Finally, the method of simplification for combined complex manufacturing systems enables computing the productivity rate for the automated line with serial structure segmented on sections with buffers. Application of such a method enables avoiding cumbersome block equations of the productivity rate for complex manufacturing systems with mixed structures and providing practical ways to use simple equations of productivity rates. This method of simplification allows for engineers to more accurately use equations and interpret results.

Productivity theory for industrial engineering enables solving basic engineering problems related to technology, design, structure and exploitation of manufacturing systems with different complexity and predict the output of systems. The principles of productivity theory for the manufacturing machine and system, formulated by mathematical models enable for identification of their characteristics, to understand their differences, regularities and reasonable links, evolution and future trends in technology and designs. These problems are alleviated by the fact that their mathematical models can unmistakably identify such characteristics. It is essential that such models of productivity rate are able to describe the production process.

Based on analytical models, productivity theory is a powerful tool, and it is possible to identify the manufacturing systems are suited for different production processes.

The principles of productivity theory for industrial engineering are as follows:

- Production processes is implemented in space and time. The productive time is separated on manufacturing time and non-productive time that is lost time for economics where 'time is money'.
- Technological processes are the basis of design for manufacturing machines and systems and determine the quality and quantity of manufacturing processes.
- Productivity rate of any manufacturing machinery is limited by physics of technology. To increase productivity rate is necessary to create new technology with high potential attributes that realised on the equipment with new design.
- Decomposition and balancing of technological processes on serial manufacturing machine tools or workstations enable an increase in the productivity rate.
- Manufacturing systems are represented by serial, parallel and parallel–serial structures with buffers that enable an increase in the productivity rate.
- Intensification of machining regimes until defined limit enables increase in the productivity rate.
- Mathematical models for the productivity rate of manufacturing machines and systems with different construction based on parameters of technology, reliability, structure and management service.

- Mathematical models for the productivity rate of manufacturing machines and systems enable finding optimal machining regimes and optimal structure for manufacturing systems by criterion of maximal productivity rate

The productivity theory of industrial engineering is based on analysis of production time for an industrial system. Theory enables the evaluation of time components and their influence on the output of the production systems. The derived mathematical models facilitate the calculation of productivity rate for the given technology, design and reliability of industrial machines and systems, as well as the management factor, which enable monitoring production systems. Each of these factor has own impact on the output of the production system. The mathematical models enable for the calculation of changes in the output of production systems based on changes in these factors, across all industrial machines and systems. The results of these calculations are shown in diagrams, which depict the change in technological productivity, design and reliability factors versus the change in design complexity of the machine and systems.

Designers of manufacturing systems can use mathematical models for calculating the productivity at the project stage of production system designs and at time of exploitation to get best engineering solution with an eye towards economical perfection of manufacturing systems.

The productivity theory represents several mathematical models that are available for calculating the productivity rate for industrial machines and systems without assumptions and simplifications. These models are accurate than the one known one in literature based on initial data of manufacturing processes. Mathematical models for productivity rate represent a set of parameters of the manufacturing systems that enable for demonstrating differences and similarities in structural solutions. Detailed presentation of these parameters in models for productivity rate surpasses all the others models and form rather a coherent entity in which differences in models are fairly large. The differences play a significant role, and optional solutions can be recommended for economic reasons. A uniform understanding prevails for the phenomenon of productivity rate for manufacturing systems and how it should be mathematically modelled and evaluated. The different mathematical models of productivity rate for different production systems reveal and describe their structural diversity that give predictable outputs. Table 7.1 represents the mathematical models for productivity rates of all types manufacturing system arrangements, automated line structures and designs and single manufacturing machine. The components of mathematical models are as specified earlier in nomenclature.

Table 7.1 Basic mathematical models for productivity rates of manufacturing systems arrangements, automated lines structures and designs and single manufacturing machine

Manufacturing system		Equation of productivity rate
Single machine	Machine tool, workstation	$Q = \dfrac{1}{t_m + t_a} \times \dfrac{1}{1 + m_r \lambda_s}$
Parallel arrangement	Independent workstations	$Q = \dfrac{p}{t_m + t_a} \times \dfrac{1}{1 + m_r \lambda_s}$
	Automated line	$Q = \dfrac{p}{t_m + t_a} \times \dfrac{1}{1 + m_r p_s \lambda_s}$
	Rotor-type machine	$Q = \dfrac{p}{(t_m + t_a)\left(2 + \dfrac{p_\gamma - 1}{p_s - p_\gamma}\right)} \times \dfrac{1}{1 + m_r p_s \lambda_s}$
Serial arrangement	Independent workstations	$Q = \dfrac{1}{(t_{mo}/q)f_c + t_a} \times \dfrac{1}{1 + m_r \lambda_{s.b}}$
	Automated line	$Q = \dfrac{1}{(t_{mo}/q)f_c + t_a} \times \dfrac{1}{1 + m_r \left(\displaystyle\sum_{i=1}^{q} \lambda_s + \lambda_{tr} + \lambda_{cs}\right)}$
	Automated line segmented on sections with buffers	$Q = \dfrac{1}{\dfrac{t_{mo}}{q} f_c + t_a} \times$ $\dfrac{1}{1 + m_r \left[\dfrac{f_s \displaystyle\sum_{i=1}^{q} \lambda_{s.i}}{n} + \Delta\lambda_{f(i-k)} + \cdots + \Delta\lambda_{f(i-2)} + \Delta\lambda_{f(i-1)} + \Delta\lambda_{b(i+1)} + \Delta\lambda_{b(i+2)} + \cdots + \Delta\lambda_{b(i+n)} + \lambda_{bf} + \lambda_v + \lambda_{tr}\right]}$

(Continued)

Table 7.1 (Continued) Basic mathematical models for productivity rates of manufacturing systems arrangements, automated lines structures and designs and single manufacturing machine

Manufacturing system		Equation of productivity rate
Parallel–serial arrangement	Independent workstations	$Q = \dfrac{p}{(t_{mo}/q)f_c + t_a} \times \dfrac{1}{1 + m_r \lambda_{s.b}}$
	Automated line of parallel structure with independent serial constructions	$Q = \dfrac{p}{(t_{mo}/q)f_c + t_a} \times \dfrac{1}{1 + m_r(p_s \lambda_{s.b} + \lambda_{cs})}$
	Automated line of serial structure with independent parallel constructions	$Q_3 = \dfrac{p}{(t_{mo}/q)f_c + t_a} \times$ $\dfrac{1}{1 + m_r\left(f_s \displaystyle\sum_{i=1}^{q} \lambda_{s.i} + \lambda_{tr} + \lambda_{cs} \right)}$
	Automated line of parallel–serial structure	$Q_3 = \dfrac{p}{(t_{mo}/q)f_c + t_a} \times$ $\dfrac{1}{1 + m_r\left(p_s \displaystyle\sum_{i=1}^{q} \lambda_{s.i} + \lambda_b + \lambda_{tr} + \lambda_{cs} \right)}$
	Automated line of parallel–serial structure segmented on sections with buffers	$Q = \dfrac{p}{\dfrac{t_{mo}}{q}f_c + t_a} \times$ $\dfrac{1}{1 + m_r\left[\begin{array}{l} \dfrac{pf_s\displaystyle\sum_{i=1}^{q}\lambda_{s.i}}{n} + \Delta\lambda_{f(i-k)} + \cdots + \Delta\lambda_{f(i-2)} \\ +\Delta\lambda_{f(i-1)} + \Delta\lambda_{b(i+1)} + \Delta\lambda_{b(i+2)} + \cdots \\ +\Delta\lambda_{b(i+n)} + \lambda_{bf} + \lambda_c + \lambda_{tr} \end{array} \right]}$

(Continued)

Table 7.1 (Continued) Basic mathematical models for productivity rates of
manufacturing systems arrangements, automated lines structures and designs
and single manufacturing machine

Manufacturing system	Equation of productivity rate
Rotor-type automated line	$$Q_{pq} = \frac{p}{\left(\dfrac{t_{mo}}{q} f_c + t_a\right)\left(2 + \dfrac{p_\gamma + p_t - 1}{p_s - p_\gamma}\right)}$$ $$\times \frac{1}{1 + m_r\left[p_s\left(\displaystyle\sum_{i=1}^{q}\lambda_{s.i} + \lambda_t\right) + \lambda_{cs}\right]}$$
Rotor-type automated line segmented on sections with buffers	$$Q_{pq} = \frac{p}{\left(\dfrac{t_{mo}}{q} f_c + t_a\right)\left(2 + \dfrac{p_\gamma + p_t - 1}{p_s - p_\gamma}\right)}$$ $$\times \frac{1}{1 + m_r\left[\begin{array}{l}\dfrac{pf_s\displaystyle\sum_{i=1}^{q}\lambda_{s.i}}{n} + \Delta\lambda_{f(i-k)} + \cdots + \Delta\lambda_{f(i-2)} \\ + \Delta\lambda_{f(i-1)} + \Delta\lambda_{b(i+1)} + \Delta\lambda_{b(i+2)} + \cdots \\ + \Delta\lambda_{b(i+n)} + \lambda_{bf} + \lambda_c + \lambda_{tr}\end{array}\right]}$$

The basic mathematical models of productivity rate (Table 7.1) enable
for getting a new point of view and possibility to analyse growth in the
productivity, defining the optimal structures and optimal machining
regimes for any of manufacturing systems to get maximal productivity
rate. Mathematical models of productivity rates reveal that technology
and design complexity of machinery are primary factors. The latter one
has an effect on the cycle time of the manufacturing process and on the
availability of industrial machines and systems. The mathematical models were tested and validated on several types of manufacturing systems.
The contribution of the productivity theory is to provide new holistic
mathematical models for the productivity rate of the manufacturing systems with a complex structure, based on the technological and technical
data that have yet to be solved in engineering.

Finally, productivity theory for industrial engineering is universal and can be applied for any type of industries and processes, i.e. for
manufacturing, mining, transport, food, chemical, agricultural, etc.
industries. Each of these industries described by own technologies, used
own machines and systems, but all of them are based on repeatable cycle

processes with different durations. It means that their processes can be described analytically in terms of productivity theory. The peculiarities of technological processes of different industries should be represented and expressed by own symbols and terms in mathematical models for productivity rates. To derive analytical basis for processes of some industries is a challenged problem of experts.

Productivity theory for industrial engineering is a powerful tool for economic development of manufacturing companies. This theory with mathematical models enables solving problems of increase in the productivity rate for production systems, defines optimal machining processes and structural designs of the complex manufacturing systems by criterion of the maximal productivity rates. Nevertheless, manufacturers are looking for solutions to fabricate the products with minimal cost and high quality. This basic economics problem combines all components of production system linked with interdependent parameters, which are the productivity rate, structural designs of manufacturing systems, managerial and service system of production processes, etc. Mathematical optimisation of interrelated multi-parametric components of a production system by criterion minimal cost is crucial problem for manufacturers. This economic problem should be solved by the theory of mathematical modelling of different production systems that enable optimising the structural designs of complex manufacturing lines, machining regimes, maximising productivity rates and managerial system, etc. by criterion of the minimal production cost for fabricated products. The known economic theory is solving partial and scattered problems of the production systems. However, the known economic models in engineering enable developing the holistic economic theory with mathematical models that combine interrelated multi-parametric components of production systems. The productivity theory for industrial engineering with mathematical models for economic optimisation of production processes combined in one textbook represent the perfect and complete holistic tool for solving all production problems.

Index